SPPU New Syllabus

CH-201

INORGANIC CHEMISTRY

[2 Credits]
Chemistry : Paper-I

For
First Year B.Sc. Semester-II
As Per New Syllabus of CBCS Pattern
From June - 2019

Dr. S. S. Jadhav
Principal,
New Arts, Commerce and Science College,
Shevgaon.

Dr. A. V. Nagawade
Vice-Principal,
Reader in Chemistry,
Ahmednagar College,
Ahmednagar.

Dr. G. S. Gugale
Head,
Department of Chemistry,
Haribhai V. Desai College,
PUNE.

AF897375

NIRALI PRAKASHAN
ADVANCEMENT OF KNOWLEDGE

N5028

F.Y.B.Sc. Inorganic Chemistry **ISBN 978-93-89686-70-8**

First Edition : December 2019

Published By:

NIRALI PRAKASHAN

Abhyudaya Pragati, 1312, Shivaji Nagar
Off J.M. Road, PUNE – 411005
Tel - (020) 25512336/37/39, Fax - (020) 25511379
Email : niralipune@pragationline.com

➢ DISTRIBUTION CENTRES

PUNE

Nirali Prakashan (For orders within Pune) : 119, Budhwar Peth, Jogeshwari Mandir Lane, Pune 411002, Maharashtra, Tel : (020) 2445 2044, Mobile : 9657703145
Email : niralilocal@pragationline.com

Nirali Prakashan (For orders outside Pune) : S. No. 28/27, Dhayari, Near Asian College Pune 411041
Tel : (020) 24690204; Mobile : 9657703143
Email : bookorder@pragationline.com

MUMBAI

Nirali Prakashan : 385, S.V.P. Road, Rasdhara Co-op. Hsg. Society Ltd., Girgaum, Mumbai 400004, Maharashtra;
Mobile : 9320129587 Tel : (022) 2385 6339 / 2386 9976, Fax : (022) 2386 9976
Email : niralimumbai@pragationline.com

➢ DISTRIBUTION BRANCHES

JALGAON

Nirali Prakashan : 34, V. V. Golani Market, Navi Peth, Jalgaon 425001, Maharashtra, Tel : (0257) 222 0395, Mob : 94234 91860;
Email : niralijalgaon@pragationline.com

KOLHAPUR

Nirali Prakashan : New Mahadvar Road, Kedar Plaza, 1st Floor Opp. IDBI Bank, Kolhapur 416 012, Maharashtra. Mob : 9850046155;
Email : niralikolhapur@pragationline.com

NAGPUR

Nirali Prakashan : Above Maratha Mandir, Shop No. 3, First Floor, Rani Jhanshi Square, Sitabuldi, Nagpur 440012, Maharashtra
Tel : (0712) 254 7129;
Email : niralinagpur@pragationline.com

DELHI

Nirali Prakashan : 4593/15, Basement, Agarwal Lane, Ansari Road, Daryaganj Near Times of India Building, New Delhi 110002
Mob : 08505972553, Email : niralidelhi@pragationline.com

BENGALURU

Nirali Prakashan : Maitri Ground Floor, Jaya Apartments, No. 99, 6th Cross, 6th Main, Malleswaram, Bengaluru 560003, Karnataka;
Mob : 9449043034
Email: niralibangalore@pragationline.com

Other Branches : Hyderabad, Chennai

niralipune@pragationline.com | www.pragationline.com

Also find us on f www.facebook.com/niralibooks

Preface ...

It gives us immense pleasure to place this text book of F.Y.B.Sc. Inorganic Chemistry in the hands of F.Y.B.Sc. students. This book has been presented to you according to the revised syllabus of Savitribai Phule Pune University implemented from June 2019.

An attempt is made to give more emphasis on essential modern concepts in chemistry. The concepts are illustrated and elaborated with necessary examples. Sufficient exercise is given at the end of each topic for the practice of the students.

Some matter in this book may not be in syllabus, but is given for the deep clarification to the students.

We hope that the book will fulfil the expectations of the students and teachers. We hope that this book will cater the exact requirements of the students.

We are grateful to our dynamic publisher Shri. Dineshbhai Furia, Shri. Jignesh Furia, and staff of Nirali Prakashan especially Mr. Malik Shaikh, Mr. Kiran Velankar, Mrs. Anjali Muley, for their efforts and keen interest in publishing this book in a very limited span of time.

Any comment, criticism and suggestions from the readers for improving the book will be highly appreciated.

Authors

Syllabus ...

1. Atomic Structure (14 L)

Origin of Quantum Mechanics: Why study Quantum Mechanics? Quantum Mechanics arose out of interplay of experiments and theory, Energy Quantization: (i) Black body radiation, (ii) The photoelectric effect, (iii) Wave particle duality: (a) The particle character of electromagnetic radiation, (b) The wave character of particle, (iv) Diffraction by double slit, (v) Atomic spectra, Review of Bohr's theory and its limitations, Heisenberg's uncertainty principle.

Quantum Mechanics: Time independent Schrödinger's equation and meaning of various terms in it, Significance of ψ and ψ^2, Schrödinger's equation for hydrogen atom. Radial and angular parts of the hydrogenic wave functions (atomic orbitals) and their variations for 1s, 2s, 2p, 3s, 3p and 3d orbitals (only graphical representation). Radial and angular nodes and their significance. Radial distribution functions and the concept of the most probable distance with special reference to 1s and 2s atomic orbitals. Significance of quantum numbers, orbital angular momentum and quantum numbers m_l and m_s. Shapes of s, p and d atomic orbitals, nodal planes. Discovery of spin, spin quantum number (s) and magnetic spin quantum number (m_s).

2. Periodic Table and Periodicity of Elements (10 L)

Periodic Table: Periodic table after 150 years, Review on the eve of International Year of Periodic Table [IYPT].

Periodicity of Elements: Rules for filling electrons in various orbitals, Electronic configurations of atoms. Stability of half filled and completely filled orbitals, Concept of exchange energy. Relative energies of atomic orbitals, Anomalous electronic configurations.

Long form of the periodic table : s, p, d and f block elements.

Detailed discussion of following properties of elements with reference to s and p block :

(a) Effective nuclear charge, shielding or screening effect

(b) Atomic and ionic radii

(c) Crystal radii

(d) Covalent radii

(e) Ionization energies

(f) Electronegativity, Pauling's / electronegativity scale

(g) Oxidation states of elements

3. Chemical Bonding (12 L)

Attainment of stable electronic configurations, Types of chemical bonds: Ionic, covalent, coordinate and metallic bonds.

Ionic Bond: General characteristics of ionic bonding, Types of ions, Energy considerations in ionic bonding, lattice energy and solvation energy and their importance in the context of stability and solubility of ionic compounds. Statement of Born-Lande equation for calculation of lattice energy, Born-Haber cycle and its applications, polarizing power and polarizability. Fajan's rules, ionic character in covalent compounds, bond moment, dipole moment and percentage ionic character.

Covalent Bond: Valence bond approach, Hybridization with suitable examples of linear, trigonal planar, square planar, tetrahedral, trigonal bipyramidal and octahedral arrangements. VSEPR theory, assumptions, need of theory, application of theory to explain geometries of molecules such as: (i) ClF_3, (ii) Cl_2O, (iii) BrF_5, (iv) XeO_3, (v) $XeOF_4$.

***

Contents ...

***

PATTERN OF UNIVERSITY EXAM QUESTION PAPER

Total Marks : 35 **Duration : 2 Hours**

Note :

(a) Q. 1 is compulsory.
(b) Solve any three questions from Q.2 to Q.5.
(c) Questions 2 to 5 carry equal marks.

1. Solve any five of the following :

(a)
(b)
(c)
(d)
(e)
(f)

(Ask four tricky questions and two questions based on problem solving, if applicable.) (5)

2. (A) Describe – type of question(s) with internal options. (6)
(B) Short question, but tricky. (4)

3. (A) Explain – type of question(s) with internal options. (6)
(B) Problem based question, if applicable otherwise justification type – question. (4)

4. (A) Discuss - type of question(s) with internal options. (6)
(B) Problem based question, if applicable otherwise tricky and signified question. (4)

5. Write short notes on any four of the following : (10)

(A) Principle based.
(B) System based.
(C) Structure based.
(D) Descriptive based.
(E) Working based.
(F) Model based.

***

Chapter 1...

Atomic Structure

Contents ...

(A) ORIGIN OF QUANTUM MECHANICS

INTRODUCTION

The most important issues in Physics had been explained successfully till the end of nineteenth century. The thermodynamics, mechanics and electrodynamics were supposed to be developed completely so that every phenomenon could be satisfactorily explained. In the beginning of the twentieth century, the development of atomic model showed some drawbacks between the results of measurements and the predictions of classical physics. The properties of microscopic particles like atom cannot be well explained with classical physics. According to classical physics, all bodies at a temperature other than zero Kelvin radiate an infinite amount of energy. The classical theory cannot explain the diffraction of an electron by a crystalline solid. The intensity and wavelength characteristic of light emitted by a radiating 'black body' was also cannot be explained by classical theory. Researchers know from the experiments that atoms consist of a small positively charged nucleus surrounded by a diffuse cloud of electrons. Classical Physics, however predicts that the atom is unstable and that the electrons will spiral into the nucleus while radiating energy. Thus, in the study of development of atomic structure, the experimental observations cannot be satisfactorily explained by classical theory which motivated the researchers for the development of quantum mechanics.

1.1 WHY STUDY QUANTUM MECHANICS ?

We know that the chemistry is a molecular science. The goal of chemist is to understand macroscopic behaviour in terms of properties of individual molecules. For example, H_2 is a good fuel because the energy released in forming H_2O is much greater than that needed to break the bonds in the reactants O_2 and H_2. We also know that the classical physics was unable to explain that why the atomic structure of positively charged nucleus surrounded by electron was stable. It is also unable to explain why graphite conduct electricity and diamond does not. The light by hydrogen discharge lamp appears at only small number of wavelengths and why the bond angle in H_2O is different from that H_2S cannot be explained by classical physics. Thus, all these phenomena which cannot be understood by classical physics can be very well explained by quantum mechanics. The central feature that distinguishes quantum and classical mechanics is wave-particle duality. At the atomic level, electrons, protons and light all behave as waves or particles. We can experimentally prove that whether the microscopic particle can behave as wave or as particle.

Now a days, quantum is part and parcel of our scientific world. The atom with central positively charged nucleus and surrounded by negatively charged electron cloud, is accepted as a stable system. The integrated circuits in computers, the laser in CD player, the chemical bond, the infrared spectroscopy, used to identify chemical compounds, the nuclear magnetic resonance spectroscopy, all these techniques would not be possible if atoms and molecules could have any value of energy as predicted by classical physics. The quantum mechanics predicts that atoms and molecules can only have discrete energies and thereby provides a basis for understanding all the above facts.

Technology is increasingly based on quantum mechanics. The quantum computing in which a state can be described by zero and one rather than zero or one is a very active area of research. If the quantum computers can ultimately be realized, they will be much more powerful than current computers. The chemical properties of biologically important molecules if calculated by using quantum mechanics are found to be sufficiently accurate so that the molecules can be designed for a specific application before testing in the laboratory. The biological science is increasingly focussed on the molecular level, therefore more scientists will need to be able to think in terms of quantum mechanical model. All this discussion indicates the need to study quantum mechanics.

1.2 QUANTUM MECHANICS AROSE OUT OF THE INTERPLAY OF EXPERIMENTS AND THEORY

The important feature of validating theories as useful model is to compare the result of a new experiment with the prediction of currently accepted theories. If the experiment and the theory agree, we gain confidence in the model underlying the theory, if not, the model needs to be modified. The atomic model developed by J. J. Thomson by performing the experiments have some drawback, therefore it was needed to modify the model. The drawbacks like atomic spectra and the radiating electron could not be explained satisfactorily by classical physics. Many scientists of that era believed that there was little left in the natural sciences that was unknown. However, the growing ability of scientists to prove natural phenomena at an atomic level soon showed that this was incorrect. The field of quantum mechanics arose in the early 1900s as scientists became able to investigate natural phenomena at this newly accessible atomic level. A number of key experiments showed that the predictions of classical physics were inconsistence with experimental outcomes. There are several experiments having important role in stimulating the development of theories to describe the natural world. These experiments stimulated the leading scientist of the era to formulate quantum mechanics.

The energy of the atomic level is continuous according to classical physics but quantum physics says that energy comes in discrete packets called quanta. The second key property is wave-particle duality. At the atomic level, light wave have particle like properties and atoms as well as sub-atomic particles such as electrons have wave like properties.

1.3 BLACK BODY RADIATION

It is the general observation that when radiation strikes any surface of a body then some part of that radiant energy is reflected, some part is absorbed and remaining part is transmitted.

The incomplete absorption of radiant energy is because of the ordinary bodies are as a rule imperfect absorbers of radiation. But in contrast to these, we have the black body, which by definition is the perfect absorber of energy and retains any radiant energy that strikes it. The black body is an hollow sphere blackened on the inside and with a small opening. (Refer Fig. 1.1).

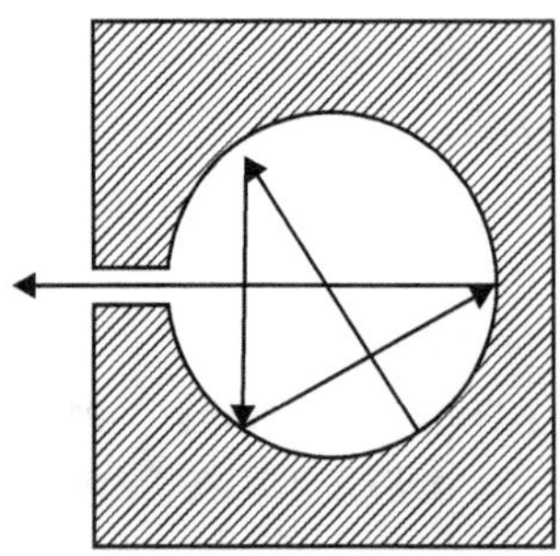

Fig. 1.1: An idealised black body

Any radiation that enters through the small opening is reflected repeatedly from the walls of the encloser until all of the energy eventually becomes absorbed. A black body is not only a perfect absorber of radiant energy, but also a perfect radiator. In fact, all bodies are heated to a given temperature, it is the black body which will radiate the maximum amount of energy possible for the given temperature. Further the black body is in thermal equilibrium with its surroundings and radiates in any given time per unit area the same amount of energy as it absorbs.

The total amount of energy 'E' radiated by a black body per unit area and time is given by the Stefan-Boltzmann fourth power law.

$$E = \sigma T^4 \qquad \dots (1.1)$$

where T is the absolute temperature, while σ is the universal constant equal to 5.6697×10^{-5} for energy in ergs and time in seconds and area in square centimeters. This energy is not emitted with a single frequency nor is it uniformly distributed along the spectrum. The energy emitted depends on temperature and wavelength. For each temperature, there is a wavelength at which the energy radiated is a maximum.

It is a common observation that, a hot object emits electromagnetic radiations. At high temperature an appreciable proportion of the radiation is in the visible region of the spectrum. The higher proportion of short wavelength, blue, light is generated as the temperature is raised. This is seen when an object glowing red hot becomes white hot when heated further. The precise depends is shown in Fig. 1.2, which shows how the energy output depends on the wavelength at several temperatures. This illustration shows two main features.

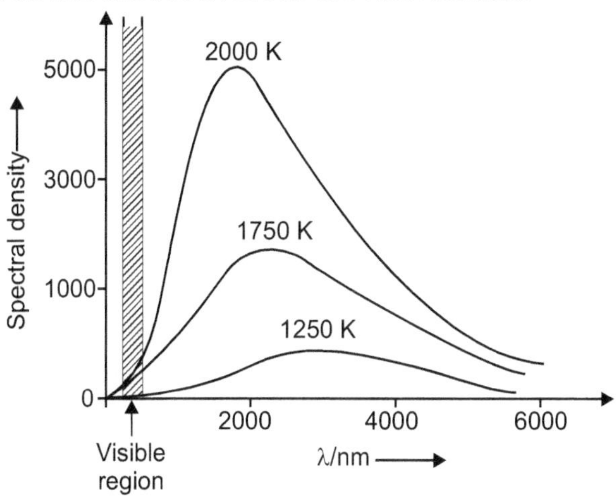

Fig. 1.2: The energy per unit volume per unit wavelength range in a black body cavity at several temperatures

The first is that the peak shifts to the shorter wavelengths as the temperature is raised and the short wavelength tail spreads through the whole of the visible region. The analysis of the data led **Wilhelm Wien** to the conclusion that the wavelength corresponding to the emission maximum is related to temperature by

Wien's Displacement Law : $T\lambda_{max}$ = Constant ... (1.2)

The experimental value of the constant is 2.9×10^{-3} mK and so at 1000 K, λ_{max} = 2900 nm.

Lord Rayleigh, with minor help from James Jeans, took a classical view and thought of the electromagnetic field as a collection of harmonic oscillator, one for each possible frequency of light. Then the occurrence of light of some frequency 'ν' was interpreted as the excitation of the oscillator of that frequency. The average energy of oscillator at temperature T is kT. Therefore, the energy density in the range of wavelengths λ to λ + dλ is the number of oscillators per unit volume in that range, dN (λ), multiplied by the average energy kT.

Hence, $du(\lambda) = kT\, dN(\lambda)$

Rayleigh-Jean's law : $du(\lambda) = \rho(\lambda)\, d\lambda$ ...(1.3)

$$\rho(\lambda) = 8\pi\, kT/\lambda^4$$

The factor 'ρ' is the energy per unit volume per unit wavelength and is called the density of states. This means that when ρ is multiplied by the wavelength range, we get the energy density, du (λ) in the range of wavelengths from λ to λ + dλ. When the energy density is multiplied by the volume of the region, we get the total energy in the region due to radiation in that wavelength range. Thus, oscillators of extremely short wavelength are strongly excited even at room temperature. According to classical physics, objects should glow in the dark. The absurd result is the ultraviolet catastrophe.

Thus, Wien in 1896 and Lord Rayleigh in 1900 made attempts to account for this distribution of black body radiation on the basis of classical concepts of continuous emission of radiation. Although Wien's equation proved fairly satisfactory at low wavelengths and Rayleigh at high ones, neither theory could account completely for the observed phenomena. This failure of classical theories of radiation led Max Planck in 1900 to discard these and give the bold hypothesis that black bodies radiate energy not continuously, but discontinuously in energy packets called quanta, given by the relation

$$E = h\nu \quad \text{... (1.4)}$$

where, E is the quantum of energy radiated, ν is the frequency, h called Planck constant, is universal constant equal to 6.6256×10^{-27} erg-sec. Thus, the equation (1.4) is the fundamental equation of the quantum theory of radiation. Planck considered the black body to consist of oscillators of molecular dimensions, each with a fundamental vibration frequency ν, and that each oscillator could emit energy either in the unit quantum $E_1 = h\nu$ or in whole number multiples n thereof, $E_n = nh\nu$. Thus, he developed a equation for energy E_λ radiated by a black body at any wavelength λ by the relation

$$E_\lambda = \frac{8\pi hc}{\lambda^5} \frac{1}{e^{hc/kT\lambda} - 1} \quad \text{... (1.5)}$$

where, c is the velocity of light, T is absolute temperature, k is the gas constant per molecule i.e. R/N. This equation have given excellent spectral distribution of energy. This equation is also reduced as well for low wavelengths to Wien's equation and high wavelengths to the Rayleigh equation.

1.4 PHOTOELECTRIC EFFECT

The photoelectric effect can be explained with the help of simple experiment. In this technique, a copper plate in a vacuum is irradiated with light radiation. This light incident on copper plate can be absorbed leading to the excitation of electrons to unoccupied energy levels. These electrons may gain sufficient energy such that some of them leave the metal and are ejected into the vacuum. If another electrode, called the collector in the vacuum system is added, the electrons that have been emitted from the copper upon illumination can be collected as shown in Fig. 1.3. This is called the photoelectric effect.

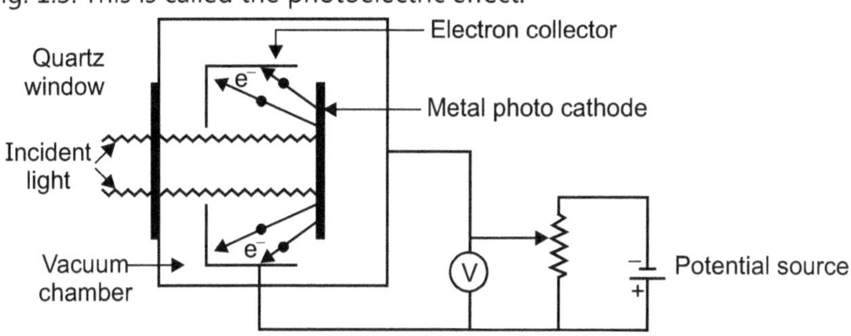

Fig. 1.3 : The electron emitted by the surface upon illumination are incident on the collector, which is at an appropriate electrical potential to attract them

By the law of conservation of energy, the absorbed light energy must be balanced by the kinetic energy of the emitted electrons. The energy required to eject an electron at equilibrium, because the energy of system is constant. The characteristics of photo-electric effect are as follows :

1. No electrons are ejected, regardless of the intensity of the light, unless its frequency exceeds a threshold value characteristic of the metal.
2. Light is incident as a phase wave over the whole copper plate. Therefore, the light is absorbed by many electrons in the solid. Any one electron can absorb only a small fraction of the incident light.
3. Electrons are emitted to the collector for all light frequencies provided that the light is sufficiently intense.
4. The kinetic energy of the ejected electron is linearly proportional to the frequency of the incident light.
5. Even at low light intensities, electrons are ejected immediately if the frequency is above threshold.

The results of the experiment can be summarised as follows :

1. The number of emitted electrons is proportional to the light intensity, but their kinetic energy is independent of the light intensity.
2. No electrons are emitted unless the frequency 'ν' is above a threshold frequency ν_o even for high light intensities.
3. The kinetic energy of the emitted electron depends on the frequency in the manner shown in Fig. 1.4.
4. Electrons are emitted even at such low light intensities that all the light absorbed by the entire copper plate is barely enough to eject a single electron, based on energy conservation considerations.

In 1905, Albert Einstein hypothesized that the energy of the light was proportional to its frequency.

$$E = \beta\nu \qquad \text{... (1.6)}$$

where, β is a constant to be determined. Thus, there is remarkable difference between classical electrodynamics in which there is no relation between the energy of light wave and its frequency. From the conservation of energy, the energy of electron E_e is related to that of light by,

$$E_e = \beta\nu - \phi \quad \text{... (1.7)}$$

The binding energy of electron in the solid, which is same as the ionization energy of an atom, is designated by ϕ in this equation and is called the work function of the metal, the energy required to remove an electron. This equation tells us that, the kinetic energy of the photoelectron that has escaped from the solid is smaller than the photon energy by the amount with which the electron is bound to the solid.

Einstein's theory gives the prediction of the dependence of kinetic energy of the photoelectrons as a function of the light frequency. ϕ can be determined independently. The β can be obtained by plotting the data of energy of photo-ejected electrons as a function of light frequency shown in Fig. 1.4. The results shown by the line in this Fig. 1.4 not only reproduce the data very well, but they yield the result that β, the slope of the line is identical to Planck's constant, h. This result, which relates the energy of the light to its frequency is one of the most widely used equation in quantum mechanics.

$$E = h\nu \quad \text{... (1.8)}$$

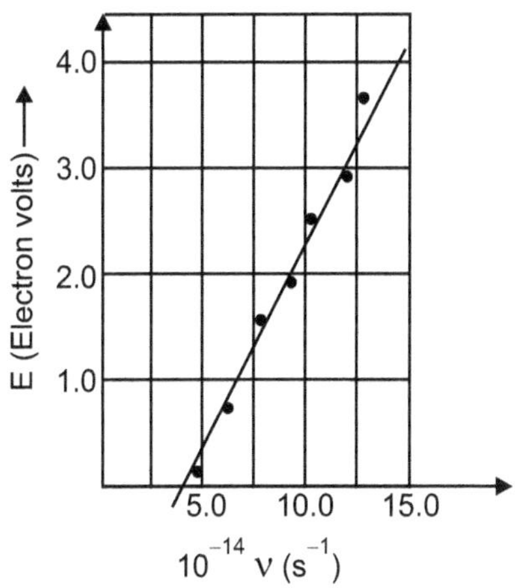

Fig. 1.4 : The energy of photo-ejected electrons is shown as a function of light frequency

The theoretical prediction and experimental data validate the fundamental assumption that Einstein made that energy of light is proportional to its frequency. This result also suggested that 'h' is a universal constant. Second important conclusion can be drawn from the observation that even at low light intensities, photoelectrons are emitted from the solid.

1.5 WAVE PARTICLE DUALITY

Dual Nature of Matter:

Wave nature of electromagnetic radiation was established at the end of 19th century. Using the wave theory, James Clerk Maxwell described successfully the light as self-propagating waves of oscillating electric and magnetic fields. Einstein's explanation of photoelectric effect (1905) showed that light has some properties of particles. Light consists of photons, each with discrete energy, $E_{photon} = h\nu_{photon}$. Thus, complete description of light includes both, wave like and particle like properties, referred to as dual nature. In wave nature delocalization is essential feature whereas in particle nature, localization is the main quality.

In 1924, French scientist Louis de-Broglie proposed that if light can display wave particle duality, then matter whose particle nature was established, might also display wave like properties under certain conditions. de-Broglie had shown that it was possible to associate waves with a moving particle such as electron. He looked for an equation which could explain the wavelength characteristics of all matter. He was able to put his ideas into quantitative relationship as

$$\lambda = \frac{h}{p} \qquad \dots (1.9)$$

where h is the Planck's constant, p is the momentum of electron, λ is the wavelength associated with the electron of mass m, moving with velocity v and the momentum p is given as p = mv. So the expression for wavelength is as shown below.

$$\lambda = \frac{h}{m - v} \qquad \dots (1.10)$$

This formula shows that a particle of smaller velocity and smaller mass is characterised by the longer wavelength. In de-Broglie's equation, electron's wavelength is a function of Planck's constant (6.625×10^{-34} joule-second) divided by electrons momentum. When this momentum is very large (relative to Planck's constant), then object's wavelength is very small. This is the case with everyday objects, the wavelength would be so small (of the order of 10^{-35} nm or smaller) as to be undetectable by any current measurement tools.

On the other hand, many small particles (such as electrons) have a very low momentum as compared to macroscopic objects. In this case, the de-Broglie's wavelength may be large enough that the particles wavelike nature gives observable effects. His equation was experimentally confirmed in 1927 when **Lester Germer** and **Clinton Davisson** who targeted electrons at a nickel crystal and the resulting diffraction pattern was found to match the predicted wavelength values.

Soon after this, George Thomson (J. J. Thomson Son) conducted an experiment (1928) and also proved the wave structure of electrons. He was bombarding thin gold foil by a beam of electrons having high velocity. While the electrons were going through the foil, the diffraction occurred. Thomson observed diffraction rings on the screen behind the foil. The two experiments described above confirmed the wave nature of the electron and de-Broglie's hypothesis was fully confirmed. So matter also has dual wave-particle nature. In no experiment we see both characteristics at the same time. Sometimes matter acts like a particle and some other time like a wave.

To understand what matter is, one must take its both characteristics into consideration. Neither the theory of particles nor the wave theory of matter are correct if considered alone. Only combining these two, one gets full and proper behaviour of microscopic matter.

In quantum mechanics, a matter wave or de-Broglie wave is the wave of matter. The de-Broglie relation shows that the wavelength is inversely proportional to the momentum of a particle and that the frequency is directly proportional to the particle's kinetic energy. The wavelength of matter is also called de-Broglie wavelength. The electromagnetic theory and hence light wave differ from de-Broglie wave in that their rest mass is zero and they travel at constant velocity in a medium, $c = 2.98 \times 10^{8}$ m/s in vacuum. de-Broglie waves can have different velocities that are much smaller than 'c'.

SOLVED EXAMPLE

Example 1.1 : *Through what potential difference an electron be accelerated so that it may have de-Broglie wavelength 0.5 A° ?*

Solution : Let V = Potential through which electron is accelerated.

$$\text{Energy} = \frac{1}{2}mv^2 = eV$$

$$\lambda = \frac{h}{(2mE)^{1/2}} = \frac{h}{(2meV)^{1/2}}$$

$$V = \frac{h^2}{2me\lambda^2}$$

Here, $h = 6.625 \times 10^{-34}$ js, $m = 9.1 \times 10^{-31}$ kg, $e = 1.6 \times 10^{-19}$ C,

$\lambda = 0.5 \times 10^{-10}$ m

Hence, $V = \mathbf{6.03 \times 10^4\ V}$ **... Ans.**

The wavelike properties in our macroscopic world do not disappear but they become insignificant. Let us consider two examples.

1. A ball of mass 140 gm is thrown with 40 m/s velocity. The de-Broglie's equation gives a wavelength of 1.2×10^{-34} m. This value is so small, actually even using the most modern technologies one would not see it.
2. On the other hand, a electron moving with 40 m/s velocity has a wavelength equal to 1.8×10^{-5} m.

The wavelike properties of particles at the atomic scale become quite significant and cannot be rejected. The magnitude of Planck's constant is so small that de-Broglie wavelength becomes significant only for small masses. Besides the wave particle duality, de-Broglie relation begins to hint at the idea of quantization. If the particle is in an orbit, the only allowed radii (and momenta) are those where the waves associated with particle will interfere non-destructively as they wrap around each orbit radii where waves destructively interfere with one another as particle orbits through successive revolutions are not allowed. Also, amplitude in de-Broglie wave reflects the particle character whereas wavelength indicates wave character.

1.6 DIFFRACTION BY A DOUBLE SLIT

The experimental study of diffraction of particles by a double slipt is carried out with particles such as neutrons. We first discuss briefly review classical diffraction of waves.

Diffraction is a phenomenon that can occur with any waves, including sound waves, water waves and electromagnetic (light) waves. Let us consider diffraction of light from a thin slit as shown in Fig. 1.5. In this experimental setup, it consists of a opaque plate containing a slit of width 'a' behind which there is a screen. The parallel rays of light are incident on the opaque plate. The screen on which the image is projected is a far away from the slit. Mathematically, this requires that b > > a.

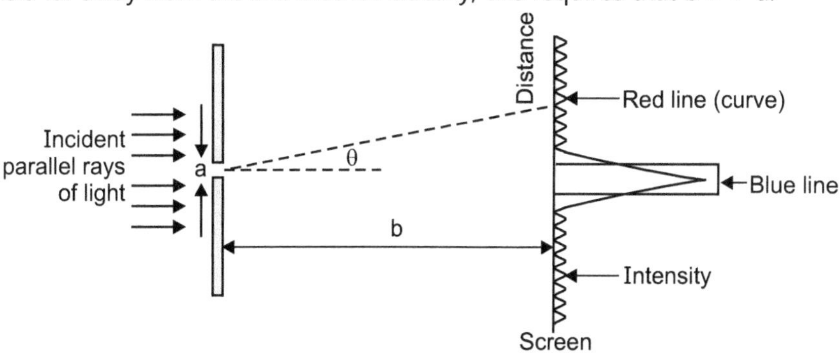

Fig. 1.5 : Diffraction of light of wavelength λ from a slit whose long axis is perpendicular to a page

When the parallel light is assumed to be incident on the slit and therefore image and slit dimensions are identical. This would give rise to an intensity pattern on the screen like that shown by blue lines in the figure, when plotted as intensity versus distance. Instead, for the condition that the light wavelength is comparable in magnitude to the slit width, an intensity distribution in the form of diffraction pattern as shown by red curve will be observed.

The origin of this pattern of alternating maxima and minima is wave interference. The maxima and minima arise as a result of path difference between the source of the cylindrical waves and the screen as shown in Fig. 1.6. The condition that the minima satisfy is

$$\sin \theta = \frac{n\lambda}{a}, \ n = \pm 1, \pm 2, \pm 3, \ldots\ldots \quad \ldots (1.11)$$

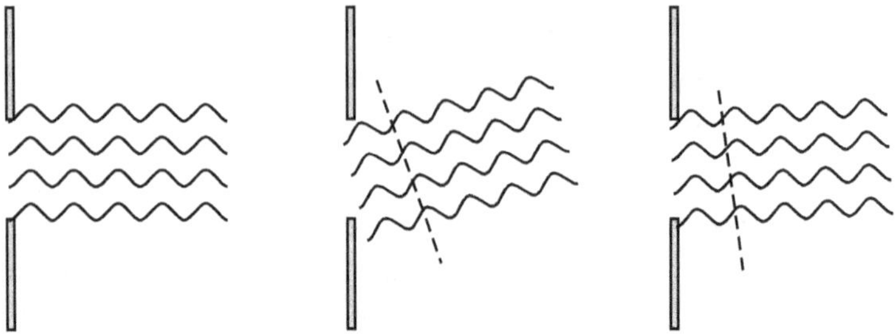

(a) The wave that emerge perpendicular to slit are all in phase and give rise to principle maxima in diffraction pattern

(b) The successive wave that emerge at the angle shown are exactly out of phase. They will interfere destructively and minimum intensity is observed

(c) Every other phase is out of phase and destructive interference with minimum intensity

Fig. 1.6

This equation helps us to understand under what conditions we might observe diffraction. The wavelength of light in the middle of the visible spectrum is about 600 nm or 6.00×10^{-4} mm. If this light is allowed to pass through a 1.00 mm wide slit and the angle calculated at which the first minima will appear, the result is $\theta = 0.03°$ for $n = 1$. This minimum is not easily observable because it lies so close to the maximum and we expect to see a sharp image of the slit on our screen. However, if the slit width is decreased to 1.00×10^{-2} mm, then $\theta = 3.4°$. This minimum is easily observable and successive bands of light and darkness will be observed instead of sharp image. The exact same behaviour is observed in wave particle duality in quantum mechanics. If the slit is much larger than the wavelength, then diffraction will not be observed.

Consider the experimental setup designed to detect the diffraction of particles shown in Fig. 1.7. It consists of a metal plate in which two rectangular slits of width 'a' have been cut. The long axis of the rectangle is perpendicular to the plane of the page. This type of setup is designed to detect diffraction from the combination of two slits.

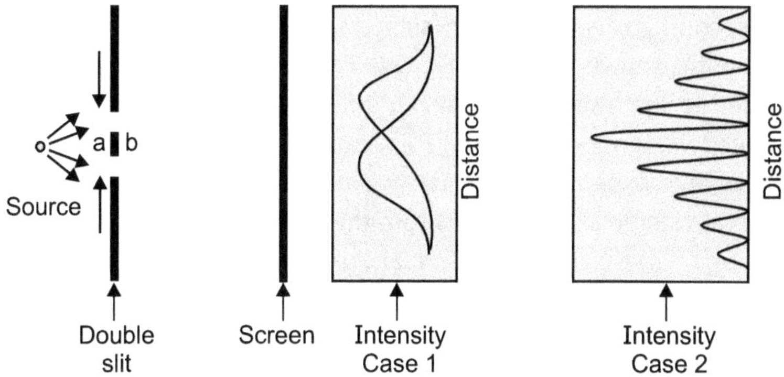

The double slit diffraction experiment

Describes the outcome of the diffraction when one of the slit is blocked

Describes the outcome when both slits are open

Fig. 1.7

We need a source of particles such as electron gun. Thus, by controlling the energy of electron the wavelength is varied. Each electron has a random phase angle with respect to every other electron. Consequently two electrons can never interfere with one another to produce a diffraction pattern. One electron gives rise to the diffraction pattern, but many electrons are needed to amplify the signal so as that we can easily see the diffraction pattern.

The screen which is placed behind the plate with slits is a phosphorescent screen that gives rise to light when incident wave or particles transfer energy to it. The electron energy is adjusted so that diffraction by single slits of width 'a' results in broad maxima. The distance between the two slits, 'b' has been chosen such that we will observe a number of intensity oscillations for small diffraction angles.

If one slit is closed off and an observer looks at the screen, he will see the broad intensity versus distance pattern shown in case 1. Refer Fig. 1.7. Depending on which slit has been closed the observer will see one of the two curves shown. The observer sees diffraction from the slit, which demonstrates that the electron can act like a wave. When working with very sensitive phosphors, the arrival of each individual electron at the screen is detected by a flash of light localized to a small area of a screen.

In second experiment both slits are kept open. The result of this experiment is shown in case 2, Fig. 1.7. For every small electron currents, the observer again sees individual light flashes localised to small areas of the screen in random pattern. This looks like particle behaviour. The results of this experiment can be interpreted as that the diffraction is seen from a single slit as well as from the double slit shows the wavelike behaviour of the electron. Individual light flashes are observed on the screen, which is due to particle trajectories rather than the diffraction pattern. The measurement of the electron current to the slit blocker seems to indicate that the electron either went through one slit or through the other. However, this conclusion is inconsistence with the appearance of a diffraction pattern because a diffraction pattern can only arise if one and the same electron goes through both slits.

1.7 ATOMIC SPECTRA

Spectroscopy is the strongest means of knowing about atomic and molecular structure and mechanics. Interaction of light with free atoms reveals information about electrons bound to individual atoms. When an atom absorbs a photon, the interaction can be of two types. If the photon has sufficiently high energy, an electron is ejected. In second type atom gains energy but does not ionise. Instead an atom is transferred to a higher energy state called an excited state. Excited state is highly unstable state. In this state they lose their excess energy in collisions with other atoms or by emitting photons in a very short span of time. Atomic energy transformations can be represented using an energy level diagram as shown in Fig. 1.8.

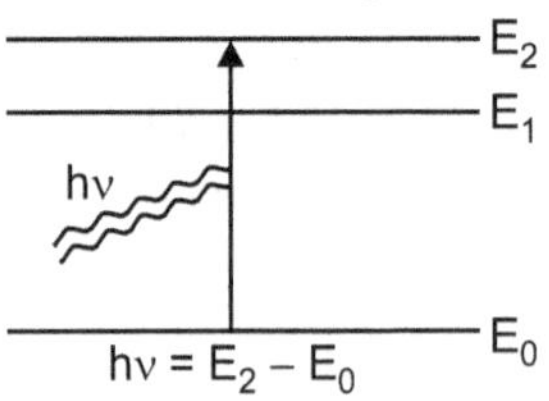

(a) Absorption of the photon causes atomic transition from E_0 to E_2

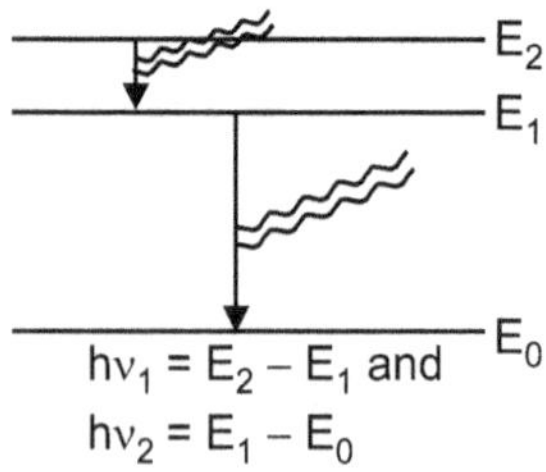

(b) Transition from E_2 to E_1 causes emission of the photon

Fig. 1.8 : Origin of the line spectra

Fig. 1.9 shows continuous spectra. A continuous spectrum can be obtained by an incandescent solid or gas. Unlike the spectrum of white light, an emission or absorption spectrum is not wide band but instead consist of narrow lines of a single colour and separated by dark bands. An emission spectrum can be produced by a gas at low pressure excited by heat or collisions with electrons. An absorption spectrum consisting of a series of dark lines results when light from a continuous source passes through a cooler gas.

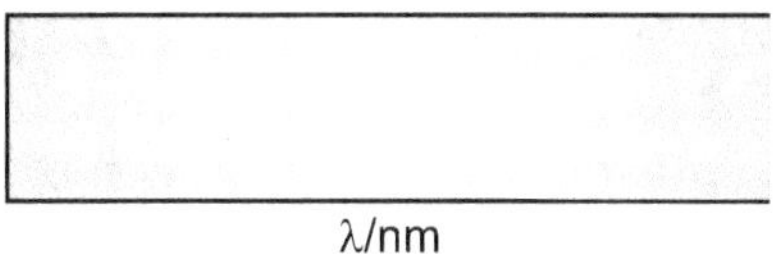

Fig. 1.9 : A continuous spectrum

If an object absorbs photon, photon energy is transferred to the object e.g. sunlight striking your skin is absorbed and warms you up. When an atom absorbs a photon, its energy increases by an amount equal to the energy of the photon. This exchange of energy is analysed in Fig. 1.10. The energy is conserved in its exchange between atom and light. This requires that change in energy of an atom exactly equals the energy of photon.

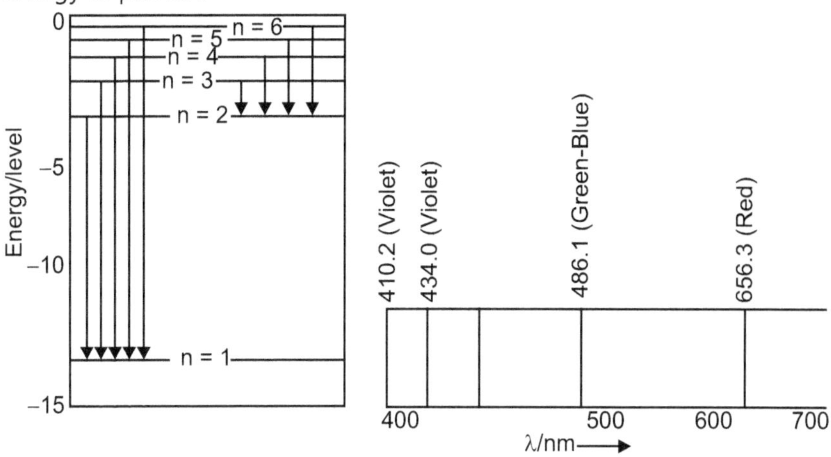

(a) Transitions of energy levels **(b) Line spectrum of an element**

Fig. 1.10

Thus, $\pm h\nu_{photon} = \Delta E_{atom}$

Each element has unique absorption and emission spectra. Photoelectric effect suggests that a photon with high enough energy can cause removal of electron from an atom. This implies that the absorption of a photon results in an energy gain for an electron in an atom and emission of photon results in loss in energy of an electron. Following mathematical description of the line spectrum of atomic hydrogen was proposed in 1885 by Johann Balmer.

$$\nu_{\text{emission}} = (3.29 \times 10^{15}\ s^{-1})\left[\frac{1}{n_1^2} - \frac{1}{n_2^2}\right] \qquad \ldots (1.12)$$

Balmer did not understand the significance of n_1 and n_2 both of which are integers. **Niels Bohr** in 1913 used the equation $E = h\nu$ to interpret Balmer's observation. The observation that the emission frequencies have specific values guided Bohr to conclude that the electron in hydrogen atom is restricted to specific energies. The integer n in equation used in describing these specific energies is referred to as principal quantum number. Thus, n_2 is the quantum number of final state and n_1 is the quantum number of initial state of the atom in above equation.

1.8 REVIEW OF BOHR'S THEORY AND ITS LIMITATIONS

The idea of smallest indivisible particle of any matter was put forth by the Greek philosopher Democritus about 450 BC. A modest English school teacher Dalton proposed Dalton's atomic theory in 1808. Most of Dalton's postulates still hold time, with few exceptions and modifications. Sir. J. J. Thomson (1897) proposed the first model of an atom in which the positive charge was imagined as being spread over the entire atom and electrons were put in this positive background. J. J. Thomson's atomic model have serious drawbacks which could not be explained by him. Ernest Rutherford in 1911 carried out an experiment i.e. α-scattering using very thin gold foil. Rutherford's atomic model was successful to explain drawbacks of Sr. J. J. Thomson's atomic model. But Rutherford's model was unsuccessful to explain stability of an atom and the nature of atomic spectra. Scientist Niel's Bohr studied these drawbacks and he modified the Rutherford's atomic model which was known as Bohr's Atomic Theory.

Postulates of Bohr's Theory :

1. The electrons revolve round the nucleus in certain fixed orbits (energy levels or stationary orbits). A certain fixed amount of energy is associated with electron in a given orbit.
2. Electrons can revolve in only those orbits for which the angular momentum is a whole number multiple of the quantity $h/2\pi$. The angular momentum is the product of mass 'm' velocity 'v' and radius r.

$$mvr = \frac{nh}{2\pi} \qquad \dots (1.13)$$

   where 'h' is Planck's constant, n is the number of orbit to which electron belongs. The above equation (1.13) is known as Bohr's quantum condition and n is known as principal quantum number.
3. As long as the electron stays in the given orbit, it does not absorb or radiate energy. The energy is absorbed or emitted only when the electron moves from the lower energy orbit to higher one or from higher energy orbit to lower one.
4. Electron absorbs or emits energy only in fixed amounts i.e. quanta. The energy change, ΔE, in electron movement can be given as,

$$\Delta E = E_2 - E_1 = h\nu \qquad \dots (1.14)$$

   where, E_1 and E_2 are initial and final energies respectively, h is Planck's constant and ν is the frequency of radiation.

1.8.1 Expression for Radius of an Electron Orbit

Consider an electron of mass 'm' and charge 'e' revolving round the nucleus having a charge (+Ze) in an orbit of radius 'r'. Let v be the tangential velocity of the electron in the orbit.

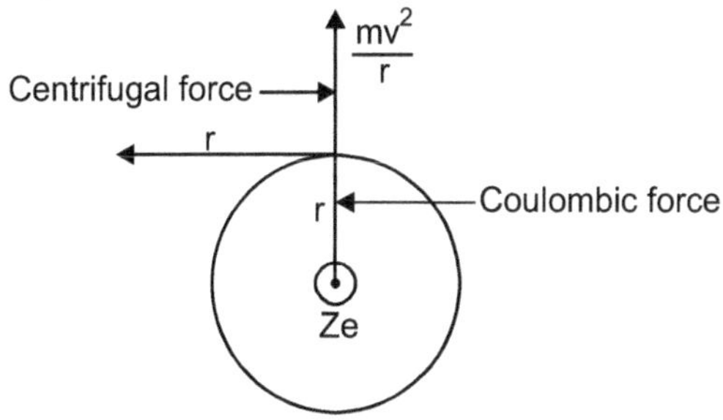

Fig. 1.11

As the electron is in stationary state, the coulombic force of attraction (acting on electron due to nucleus of charge +Ze) situated at a distance r is balanced by a centrifugal force acting on an electron due to its circular motion. Hence,

Electrostatic attractive force = Centrifugal force

i.e. $$\frac{Ze^2}{r^2} = \frac{mv^2}{r} \quad \text{... (1.15)}$$

$$\therefore \quad r = \frac{Ze^2}{mv^2} \quad \text{... (1.16)}$$

But the angular momentum (mvr) of an electron is quantized i.e. only those orbits are allowed for which angular momentum is an integral multiple of h/2π.

i.e. $$mvr = \frac{nh}{2\pi}$$

where n is the principal quantum number and n = 1, 2, 3, ...

Thus, $$v = \frac{nh}{2\pi mr} \quad \text{... (1.17)}$$

Substituting equation (1.17) into equation (1.16), we get,

$$r = \frac{Ze^2}{m\left(\frac{nh}{2\pi mr}\right)^2}$$

i.e. $$r = \frac{4\pi^2 mr^2 Ze^2}{n^2h^2}$$

$$\therefore \quad r = \frac{n^2h^2}{4\pi^2 mZe^2} \quad \text{... (1.18)}$$

The equation (1.18) shows that the radius of the orbit is directly proportional to n and since n is a whole number, only certain 'r' values are allowed.

For hydrogen $Z = 1$

$$\therefore \quad r = \frac{n^2h^2}{4\pi^2 me^2}$$

and the radius of the first orbit (n = 1) in hydrogen would be

$$r = \frac{h^2}{4\pi^2 me^2}$$

1.8.2 Expression for Velocity

Substituting equation (1.18) into equation (1.17), we get,

$$v = \frac{nh}{2\pi m\left(\frac{n^2h^2}{4r\pi^2 mZe^2}\right)}$$

$$v = \frac{2\pi Ze^2}{nh} \qquad \dots (1.19)$$

1.8.3 Expression for Energy

The total energy of an electron is the sum of the kinetic energy $\left(\frac{1}{2}mv^2\right)$ and potential energy. The potential energy arises from the work that is done on the electron in bringing it from infinity to a distance 'r' from the nucleus. It is $\left(\frac{-Ze^2}{r}\right)$. Thus,

$$\text{Total energy} = \text{Kinetic energy} + \text{Potential energy}$$

i.e.
$$E = \frac{1}{2}mv^2 - \frac{Ze^2}{r}$$

From equation (1.21), we have,

$$\frac{Ze^2}{r^2} = \frac{mv^2}{r}$$

$$\therefore \quad mv^2 = \frac{2e^2}{r}$$

$$\therefore \quad E = \frac{1}{2}\frac{Ze^2}{r} - \frac{Ze^2}{r}$$

$$\therefore \quad E = -\frac{1}{2}\frac{Ze^2}{r} \qquad \dots (1.20)$$

Substituting equation (1.18) into equation (1.20), we get

$$E = \frac{-Ze^2}{2\left(\frac{n^2h^2}{4\pi^2 mZe^2}\right)}$$

$$E = -\frac{2\pi^2 mZ^2e^4}{n^2h^2} \qquad \dots (1.21)$$

1.8.4 Application of Bohr's Theory to Atomic Spectra

If an electron jumps from higher energy level (E_{n_2}) to lower energy level (E_{n_1}), (where n_1 and n_2 are the principal quantum numbers and

$n_1 < n_2$) a photon of frequency ν is emitted. It is related to the energy difference ($E_{n_1} - E_{n_2}$) that can be obtained by using equation (1.21).

$$E_{n_1} - E_{n_2} = \Delta E = h\nu = \frac{2\pi^2 m Z^2 e^4}{h^2}\left(\frac{1}{n_1^2} - \frac{1}{n_2^2}\right)$$

$$\therefore \qquad \nu = \frac{2\pi^2 m Z^2 e^4}{h^3}\left(\frac{1}{n_1^2} - \frac{1}{n_2^2}\right) \qquad \ldots (1.22)$$

But, we know that, $\bar{\nu} = \frac{\nu}{c}$

where, $\bar{\nu}$ = Wave number

c = Velocity of light

$$\therefore \qquad \bar{\nu} = \frac{2\pi^2 m Z^2 e^4}{h^3 c}\left(\frac{1}{n_1^2} - \frac{1}{n_2^2}\right) \qquad \ldots (1.23)$$

or

$$\bar{\nu} = R_\infty \cdot Z^2 \left(\frac{1}{n_1^2} - \frac{1}{n_2^2}\right)$$

where R_∞ is a constant and is given by the equation

$$R_\infty = \frac{2\pi^2 m e^4}{h^3 c}$$

Equation (1.23) gives the wave number of the spectral lines in hydrogen spectrum. Substituting the values of m, e, h and c and Z = 1 for hydrogen, we get the theoretical expression of the hydrogen atom spectra as,

$$\bar{\nu} = \log 737 \left(\frac{1}{n_1^2} - \frac{1}{n_2^2}\right) \text{cm}^{-1}$$

When this equation is compared with empirical Rydberg's equation

$$\bar{\nu} = \log 677 \left(\frac{1}{n_1^2} - \frac{1}{n_2^2}\right) \text{cm}^{-1}$$

The value of Rydberg's constant (log 677) and the value obtained by theoretical calculations by Bohr's theory (log 737) are in good agreement. This is the remarkable achievement of Bohr's theory. Fig. 1.12 shows the transitions between various energy levels of the hydrogen atom.

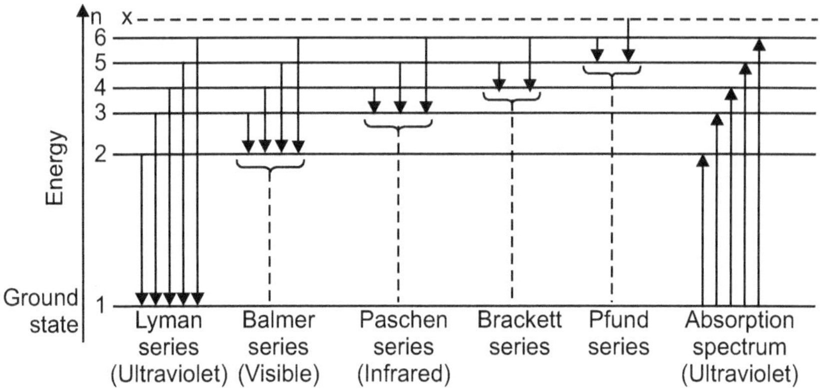

Fig. 1.12 : Transitions in hydrogen atom

1.8.5 Limitations of Bohr's Model of an Atom

Following are the limitations of Bohr's model of an atom.

1. Bohr's theory could explain the spectral details of simple atoms like hydrogen, helium etc. however it failed to explain the spectrum of elements with higher atomic number.
2. When atomic source of radiation is placed in magnetic field, a line in an emission spectrum splits into a number of closely spaced lines, called Zeeman effect. A similar splitting occurs if atomic source of radiation is kept in an electric field which is called Stark effect. Bohr's theory could not explain Zeeman and Stark effect.
3. It cannot explain the dual nature (wave as well as particle nature) of the electron that is evidenced by experiment.
4. The theory assumes that an electron in an atom is located at a definite distance from the nucleus and is associated with a fixed value of angular momentum. This is against Heisenberg's uncertainty principle, according to which it is impossible to determine simultaneously the exact position and momentum of a particle like electron.
5. No justification was given for the principle of quantization of angular momentum i.e. for the assumption that the electrons can revolve only in those orbits in which the angular momentum of an electron (mvr) is a whole number multiple of $h/2\pi$.

1.9 UNCERTAINTY PRINCIPLE

Nature has imposed limitation on human measurement of dynamic microscopic particles like electrons. **Werner Heisenberg** realised that more precisely we can define the position of an electron (a microscopic particle), less certainly we are able to define its velocity (or momentum) and vice versa. If Δx is the uncertainty in defining position and Δp the uncertainty in momentum then the uncertainty principle may be expressed mathematically as

$$\Delta x \cdot \Delta p \geq \frac{h}{4\pi} \qquad \ldots (1.24)$$

where, h is Planck's constant equal to 6.626×10^{-34} Js

The product of uncertainties in momentum and uncertainties in position cannot be less than a quantity of the order of $h/4\pi$. This means that it is impossible to know both the position and momentum (or velocity) exactly. This principle is the consequence of the wave particle duality of matter.

Note that the uncertainties in question are characteristics of system itself. In real world measurements, there can be additional uncertainties created by non-ideal and imperfect measurement process. The uncertainty principle holds true regardless of whether the measurements are ideal. Also note that the uncertainty principle has negligible effect on objects of macroscopic scale since the product of uncertainties is so small of the order of 10^{-35} joule-seconds. The explanation of the uncertainty principle depends on the wave nature of a particle. Just as it is non-sensical to discuss the precise location of a wave on a string, particles do not have perfectly precise positions, like wise just as it is non-sensical to discuss the wavelength of a 'pulse' wave travelling down a string, particles do not have perfectly precise momenta (which corresponds to the inverse of wavelength). When position is relatively well defined, the wave is pulse-like and has a very ill defined wavelength (and thus momentum). And conversely, when momentum (and thus wavelength) is relatively well defined, the wave looks long and sinusoidal and therefore it has a very ill-defined position.

(B) QUANTUM MECHANICS

1.10 TIME INDEPENDENT SCHRODINGER'S EQUATION AND MEANING OF VARIOUS TERMS IN IT

The expression of Schrodinger's equation is obtained first by knowing the term operator. An operator is a generalization of the concept of a function. A function is a rule for turning one number into another, whereas an operator is a rule for turning one function into another. For example, we consider $\frac{dy}{dx}$ to be the $\frac{d}{\delta x}$ operator operating on the function y(x). An operator is generally denoted by a capital letter with a carat (∧) over it. Thus, $\hat{A}$ f(x) = g(x) indicates that the operator $\hat{A}$ operates on function f(x) to give a new function g(x).

(A) The Hamiltonian Operator H:

The rule for setting up the quantum mechanical operator is (i) to take the classical expression for a dynamical variable, (ii) replace x, y, z and t co-ordinates and momenta p_x, p_y, p_z by their respective operators. (If the variables x, y, z, p_x, p_y and p_z occur as their n^{th} power, then corresponding operators are to be applied n times successively).

The operator for energy called the Hamiltonian operator H can thus be derived using these rules. The Hamiltonian function (in classical mechanics) is

$$H = T + V \quad \dots (1.25)$$

where, T is the kinetic energy of the system, and V is the potential energy. For a conservative system, Hamiltonian function is the energy expressed in terms of co-ordinates and conjugate momenta. If conservative system is one whose total energy is conserved with respect to time, thereby potential energy becomes the function of co-ordinates only and not that of time.

For single particle system therefore,

$$H = \frac{1}{2}mv^2 + V(x, y, z)$$

$$= \frac{m^2v^2}{2m} + V(x, y, z)$$

$$= \frac{p^2}{2m} + V(x, y, z) \text{ since, } p = mv \text{ is the momentum.}$$

$$H = \frac{1}{2m}(p_x^2 + p_y^2 + p_z^2) + V(x, y, z) \quad \dots (1.26)$$

This is the classical expression for the variable 'energy' with components of momentum as p_x, p_y, p_z and V being the potential energy.

The quantum mechanical energy operator i.e. Hamiltonian operator H is given by replacing p_x, p_y and p_z by momentum operator in above equation. Thus,

$$H = \frac{1}{2m}(p_x^2 + p_y^2 + p_z^2) + V(x, y, z)$$

$$= \frac{1}{2m}\left[\left(\frac{h}{2\pi_i}\frac{\partial}{\partial x}\right)^2 + \left(\frac{h}{2\pi_i}\frac{\partial}{\partial y}\right)^2 + \left(\frac{h}{2\pi_i}\frac{\partial}{\partial z}\right)^2\right] + V(x, y, z)$$

Since operator for momentum p_x is $\frac{h}{2\pi_i} \cdot \frac{\partial}{\partial x}$

$$H = \frac{-h^2}{8\pi^2 m}\left[\frac{\partial^2}{\partial^2 x} + \frac{\partial^2}{\partial^2 y} + \frac{\partial^2}{\partial^2 z}\right] + V(x, y, z) \qquad \because (i)^2 = -1$$

$$H = -\frac{h^2}{8\pi^2 m}\nabla^2 + V(x, y, z)$$

where, $\frac{\partial^2}{\partial^2 x} + \frac{\partial^2}{\partial^2 y} + \frac{\partial^2}{\partial^2 z} = \nabla^2$ is Laplacian operator, that represents the sum of the second partial derivatives along the three co-ordinates.

Once an operator is derived, an eigen value equation can be obtained. Thus, $\hat{A}\psi = a\psi$ is a general eigen value equation for a dynamical variable 'a'. The eigen value equation for energy as a variable of a system can be written using the energy operator Hamiltonian. Therefore, $H\psi = E\psi$

$$\left[-\frac{h^2}{8\pi^2 m}\nabla^2 + V(x, y, z)\right]\psi = E\psi$$

$$-\frac{h^2}{8\pi^2 m}\nabla^2\psi + V(x, y, z)\psi = E\psi \qquad \ldots (1.27)$$

This is called time independent Schrodinger's equation. The fundamental equation enables to study the stationary states of quantum mechanical systems. The stationary states are those states for which the values of observables are independent of time. The familiar orbitals (1s, $2p_x$ etc.) of hydrogen atom are examples of solutions $\psi(x)$ for stationary states obtained from this equation. Schrodinger's equation essentially states that, the energy E depends on two things, V and the second derivatives of ψ. Since V is a potential energy, the second derivative must be related to kinetic energy. We know that derivative is a measure of slope (of a function), therefore, the second derivative of ψ (w.r.t. a given direction) is the measure of the rate of change of slope i.e. curvature of ψ.

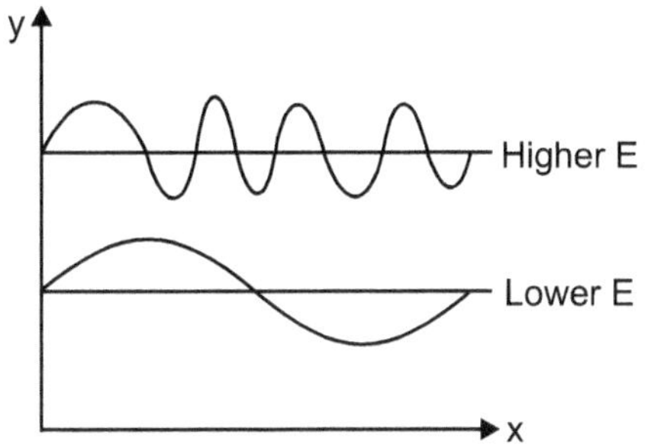

(a) Since V = 0, E = T. For higher kinetic energy, ψ is more wiggly or wavy

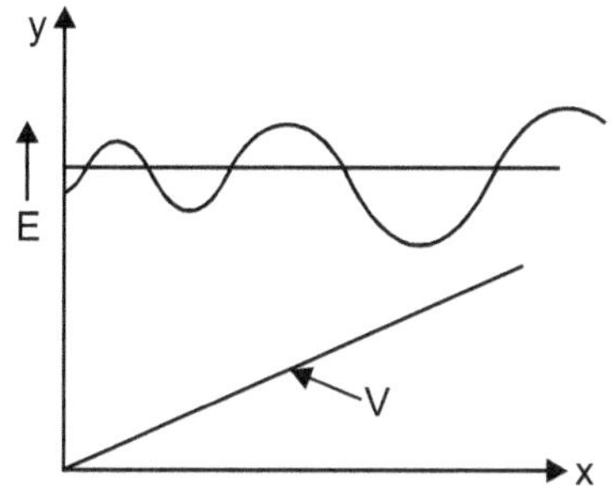

(b) ψ becomes less wiggly from left to right as V increases

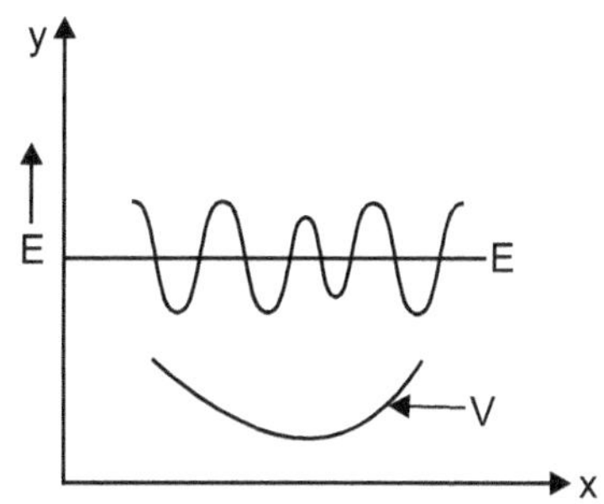

(c) ψ is less wiggly, where T is lowest and V is greatest and vice a versa

Fig. 1.13

Hence, more wiggly wave function leads to a higher kinetic energy. Since E is a constant for a state of the solutions of Schrodinger's equation must be more wiggly in regions where V is low and less wiggly, where V is high. Examples for some one dimensional cases are depicted in Fig. 1.13.

Schrodinger's equation treats electrons as waves, electrons are spread out (delocalized) rather than located in one place at a given time. Instead of fixed orbits you need to talk in terms of probabilities using quantum mechanics. Probabilities per unit volume are known as electron densities.

1.11 SIGNIFICANCE OF ψ AND ψ^2

The wave function is the variable quantity that mathematically describes the wave characteristics of a particle. The value of the wave function of a particle is related to the likelihood of the particle being at a given point in space at a given time, and may be thought of as an expression for the amplitude of the particle wave, although for such waves amplitude has no physical significance. The square of the wave function ψ^2 is the significant quantity, as it gives the probability for finding the particle at a given point in space and time. Wave function is the function of the co-ordinates of the particles of a system and of time. It determines the average result of every possible experiment on the system. It is also known as probability amplitude function or wave function or psi function. Wave function of a system can be obtained as a solution of the Schrodinger's equation for the system. Mathematically, it is a function from a space that maps the possible states of the system into the complex numbers. The laws of quantum mechanics (the Schrodinger's equation) describes how the wave function evolves over time. Thus, each particle is represented by a wave function ψ (position, time) such that $\psi\psi^*$ = Probability of finding the particle at that position at that time. ψ^* is the complex conjugate of ψ. Wave function ψ contains all the measurable information about the particle. ψ allows energy calculations through Schrodinger's equation. It permits calculations of most probable value (expectation value) of a given variable.

1.12 SCHRODINGER'S EQUATION FOR HYDROGEN ATOM

Consider an atom consisting of a nucleus of mass m and charge Ze around which revolves one electron of mass m. In mechanics, such a combination can be looked upon as a particle of reduced mass μ given by,

$$\frac{1}{\mu} = \frac{1}{m} + \frac{1}{m_i} \qquad \dots (1.28)$$

revolving about the centre of mass at a distance r from the origin. Again the potential energy of such a particle is $U = -Ze^2/r$. Inserting these values of μ and U into the Schrodinger's equation, we get,

$$\nabla^2\psi + \frac{8\pi^2\mu}{h^2}\left(E + \frac{2e^2}{r}\right)\psi = 0 \qquad \dots (1.29)$$

This equation (1.29) is best handled in spherical polar co-ordinates whose relation to Cartesian co-ordinates is shown in Fig. 1.14. In terms of r, θ and ϕ, equation (1.29) becomes as follows :

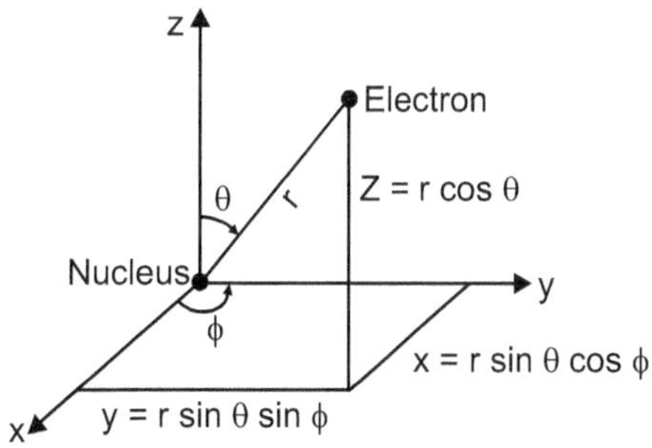

Fig. 1.14: Spherical polar co-ordinates

$$\frac{1}{r^2}\frac{\partial}{\partial r}\left(r^2\frac{\partial\psi}{\partial r}\right)+\frac{1}{r^2}\frac{1}{\sin^2\theta}\frac{\partial^2\psi}{\partial\psi^2}+\frac{1}{r^2\sin\theta}\frac{\partial}{\partial\theta}\left(\sin\theta\frac{\partial\psi}{\partial\theta}\right)+\frac{8\pi^2\mu}{h^2}\left(E+\frac{2e^2}{r}\right)\psi=0 \quad \dots (1.30)$$

$$\psi = R\cdot\Theta\cdot\Phi \quad \dots (1.31)$$

where R is a function of r only, Θ is a function of Φ only and Φ is a function of ψ. Then equation (1.30) is separated into three differential equations. All these equations have to be solved in sequence in order to obtain the allowed values of Φ, Θ and R and also of the energy E. We can state some general conclusions that emerge from the exact mathematical treatment.

(i) There are infinite number of wave functions for the electron in the hydrogen atom. This feature is alternatively expressed by the statement that there is an infinite number of orbitals for the electron in the hydrogen atom. The term 'atomic orbital' is used for an electron wave function since the wave function gives the maximum possible information about electron motion in an atom. The term 'orbital' is reminiscent of orbits of the Bohr theory but it must be remembered that orbitals do not define a definite path for the electron, rather they define only the probabilities of the electron being in various regions of the space around the nucleus. Also orbits are necessarily planar (i.e. two dimensional) but orbitals give a three dimensional distribution.

(ii) In each orbital, electron has precise energy. The magnitudes of the angular momentum, as well as one component of the angular momentum, in a reference direction are also sharp. The other two components of the angular momentum, do not have fixed values. In other words, the direction of the electron angular momentum is only partially specified.

(iii) The values of energy, the angular momentum as well as of the component of the angular momentum change discontinuously from one orbital to another. These features, first encountered in the Bohr theory, are supported by experiment. The major difference is that the magnitude and the component of angular momentum in the Bohr model were related to a precise orbit and a precise orientation of the orbit, while such an interpretation is not permissible in the Schrodinger model.

(iv) All other dynamical variables like position, velocity etc. do not have sharply defined values. For these variables, only probability distribution function exist and these can be obtained from the wave function. The square of the wave function, ψ^2, directly gives the position probability distribution for a particular orbital. ψ^2 varies from place to place and its value at a particular point measures the probability of finding the electron at that point.

(v) The permitted values, E_n, of energy (energy levels) are given by the formula $E_a = \frac{-2\pi^2 me^4}{n^2h^2}$, which is identical to that obtained by the Bohr theory. As before, n is called the principal quantum number, with only non-zero positive integral values 1, 2, 3 to ∞.

(vi) For a given value of the energy, the angular momentum cannot have any magnitude. The permitted values of the magnitude of angular momentum, L, for an energy level with quantum number n are given by the formula,

$$L = \sqrt{l(l+1)}\frac{h}{2\pi} \quad \text{... (1.32)}$$

where, l can have any positive integral value from zero to n − 1, l is known as the azimuthal quantum number.

(vii) For a particular value of the magnitude of angular momentum, there can be only certain components in a reference direction. These components have values m_l (h/2π), where m_l is an integer with any of the values from $-l$ to $+l$. m_l is an integer with any of the values from $-l$ to $+l$. m_l is known as the magnetic quantum number. l and m_l together are called the angular momentum quantum numbers. Every permitted combination of the three quantum numbers n, l and m_l defines a permitted orbital. The atomic orbitals are designated by symbols. Orbitals with $l = 0$, $l = 1$, $l = 2$ and $l = 3$ are called as s, p, d and f orbitals respectively. The principal quantum number of the orbital is given before the angular momentum symbols. Thus, the ground state orbital of hydrogen atom is written as 1s. The m_l quantum number is indicated by p_x, d_{xy}, etc.

All the above conclusions follow from the Schrodinger's equation. There is one feature of the hydrogen atom which, however, does not emerge from the equation, but is introduced in an adhoc manner to interpret experimental data. This is the concept of the electron spin. In Bohr's theory the orbital angular momentum alone was not sufficient to explain magnetic data. Thus, for the Schrodinger's theory, where the magnetic momentum calculated from the orbital angular momentum does not agree with the measured value. The assumption of spin angular momentum is made in Schrodinger's model. The magnitude of this spin angular momentum 's' is given by

$$s = \sqrt{s(s+1)}\,\frac{h}{2\pi}$$

where, $$s = \frac{1}{2} \qquad \text{... (1.33)}$$

There are only two allowed components of the spin angular momentum along a reference direction given by $\frac{1}{2}\left(\frac{h}{2\pi}\right)$ and $-\frac{1}{2}\left(\frac{h}{2\pi}\right)$. The quantum number which designates this is called the spin quantum number, m_s. It has the value $+\frac{1}{2}$ or $-\frac{1}{2}$. The spin quantum number is indicated by arrows ↑ or ↓ up or down spin.

The spin magnetic moment μ_s is given as

$$\mu_s = \frac{eh}{2\pi mc}\sqrt{s(s+1)} \qquad \text{... (1.34)}$$

To sum up then, electron behaviour in the hydrogen atom is characterised by four quantum numbers n, l, m and m_s.

1.13 RADIAL AND ANGULAR PARTS OF THE HYDROGENIC WAVE FUNCTIONS (ATOMIC ORBITALS) AND THEIR VARIATIONS FOR 1s, 2s, 2p, 3s, 3p AND 3d ORBITALS (ONLY GRAPHICAL REPRESENTATION)

The wave function for an electron in an atom is called an atomic orbital. For any electron the orbital is determined by the values of the three quantum numbers i.e. n, *l* and m. For hydrogen like atom the total wave function is the product of the radial contribution to ψ, namely R, and the two angular portions Θ and Φ. The expression for R and Θ, Φ deduced from solution of the equation (1.30) for values of n = 1 to 3 are given in Table 1.1.

Where, $$q = \frac{Z_r}{a_0}$$

and $$a_0 = \frac{h^2}{4\pi^2\mu e^2} = 0.529\ \mathring{A} \qquad \ldots (1.35)$$

The radius of the orbit is given by Bohr's equation $r = \frac{n^2h^2}{4\pi^2 mZe^2}$ for hydrogen atom. The orbitals for the hydrogen atom follow from Table 1.1 on setting Z = 1.

We first consider the radial contribution to ψ for hydrogen atom. In every moment the R^2 represents the probability of finding the electron at a point between r and r + dr in space. But it is very important to find the probability of finding the electron in a spherical shell of thickness dr at a distance r from the centre of mass, which is practically identical with the position of the nucleus. This probability D(r), called the radial distribution function is given by the expression

$$D(r) = 4\pi r^2 R^2 \qquad \ldots (1.36)$$

The plots of D(r) versus r for various values of n and *l* are shown in Fig. 1.15. For the ground state of the hydrogen atom i.e. n = 1 and *l* = 0, D(r) shows a maximum at $r = a_0 = 0.529\ \mathring{A}$, and hence the most probable position for electron is identical with the radius predicted by Bohr for first electronic orbit. Bohr predicted that the electron will be only at this distance, the plot shows definite probabilities for finding the electron at distances both shorter and longer than a_0.

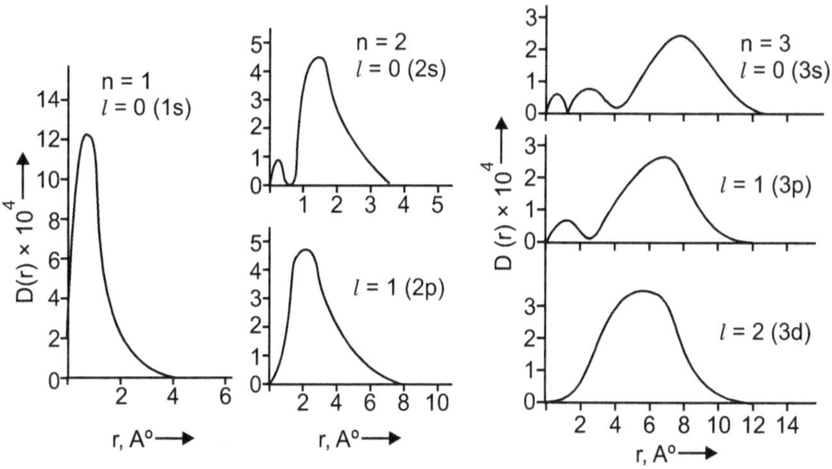

Fig. 1.15: Radial distribution functions for hydrogen atom

Table 1.1: The hydrogen like wave functions

Sr. No.	Orbital Designation	n	l	m	R	$\Theta\Phi$
1.	1s	1	0	0	$2\left(\frac{Z}{a_0}\right)^{3/2} e^{-q}$	$\left(\frac{1}{4\pi}\right)^{1/2}$
2.	2s	2	0	0	$\frac{1}{2\sqrt{2}}\left(\frac{Z}{a_0}\right)^{3/2} (2-q)\, e^{-q/2}$	$\left(\frac{1}{4\pi}\right)^{1/2}$
3.	$2p_z$	2	1	0	$\frac{1}{2\sqrt{6}}\left(\frac{Z}{a_0}\right)^{3/2} q\, e^{-q/2}$	$\left(\frac{3}{4\pi}\right)^{1/2} \cos\theta$
4.	$2p_x$	2	1	1	$\frac{1}{2\sqrt{6}}\left(\frac{Z}{a_0}\right)^{3/2} q\, e^{-q/2}$	$\left(\frac{3}{4\pi}\right)^{1/2} \sin\theta\cos\phi$
5.	$2p_y$	2	1	−1	$\frac{1}{2\sqrt{6}}\left(\frac{Z}{a_0}\right)^{3/2} q\, e^{-q/2}$	$\left(\frac{3}{4\pi}\right)^{1/2}$ $\sin\theta\sin\phi$
6.	3s	3	0	0	$\frac{2}{81\sqrt{3}}\left(\frac{Z}{a_0}\right)^{3/2}$ $(27-18q+2q^2)\, e^{-q/3}$	$\left(\frac{1}{4\pi}\right)^{1/2}$
7.	$3p_z$	3	1	0	$\frac{4}{81\sqrt{6}}\left(\frac{Z}{a_0}\right)^{3/2} (6-q)\, q\, e^{-q/3}$	$\left(\frac{3}{4\pi}\right)^{1/2} \cos\theta$
8.	$3p_x$	3	1	1	$\frac{4}{81\sqrt{6}}\left(\frac{Z}{a_0}\right)^{3/2} (6-q)\, q\, e^{-q/3}$	$\left(\frac{3}{4\pi}\right)^{1/2}$ $\sin\theta\cos\phi$

Sr. No.	Orbital Designation	n	*l*	m	R	ΘΦ
9.	$3p_y$	3	1	-1	$\frac{4}{81\sqrt{6}}\left(\frac{Z}{a_0}\right)^{3/2} (6-q)\, q\, e^{-q/3}$	$\left(\frac{3}{4\pi}\right)^{1/2}$ $\sin\theta \sin\phi$
10.	$3d_{z^2}$	3	2	0	$\frac{4}{81\sqrt{30}}\left(\frac{Z}{a_0}\right)^{3/2} q^2 e^{-q/3}$	$\left(\frac{5}{16\pi}\right)^{1/2}$ $(3\cos^2\theta - 1)$
11.	$3d_{zx}$	3	2	1	$\frac{4}{81\sqrt{30}}\left(\frac{Z}{a_0}\right)^{3/2} q^2 e^{-q/3}$	$\left(\frac{15}{4\pi}\right)^{1/2}$ $\sin\theta \cos\theta \cos\phi$
12.	$3d_{yz}$	3	2	-1	$\frac{4}{81\sqrt{30}}\left(\frac{Z}{a_0}\right)^{3/2} q^2 e^{-q/3}$	$\left(\frac{15}{4\pi}\right)^{1/2}$ $\sin\theta \cos\theta \sin\phi$
13.	$3d_{x^2-y^2}$	3	2	2	$\frac{4}{81\sqrt{30}}\left(\frac{Z}{a_0}\right)^{3/2} q^2 e^{-q/3}$	$\left(\frac{15}{16\pi}\right)^{3/2}$ $\sin^2\theta \cos 2\phi$
14.	$3d_{xy}$	3	2	-2	$\frac{4}{81\sqrt{30}}\left(\frac{Z}{a_0}\right)^{3/2} q^2 e^{-q/3}$	$\left(\frac{15}{16\pi}\right)$ $\sin^2\theta \sin 2\phi$

The radial distribution functions for n = 2 and 3 are less sharp than the one for n = 1, and they extend over wider values of r. These are the important facts which indicate a lesser localization of the electron with increase in distance from the nucleus. Therefore, there is decrease in attraction between the nucleus and the electron. For electron in 2s, 3s and 3p states, we observe relatively large maximum in D(r) accompanied by smaller maxima at lower values of r. In these states the electron has definite probabilities for occupying more than one preferred distance from the nucleus.

The neat observations of the Table 1.1 show that all the orbitals having identical values of *l* and m have the same dependence of θϕ on the two angles irrespective of the values of n and R. All the 's' orbitals will have the same shape and so will all p_z orbitals and p_x orbitals etc. From Table 1.1, for s orbitals θϕ is a constant, and hence all such orbitals will have spherical symmetry. All other types give dependence on θϕ on the angles, all orbitals other than 's' will have shapes which will not be spherically symmetrical.

Three dimensional plots of θϕ are shown in Fig. 1.16. As expected, s orbitals are spherical.

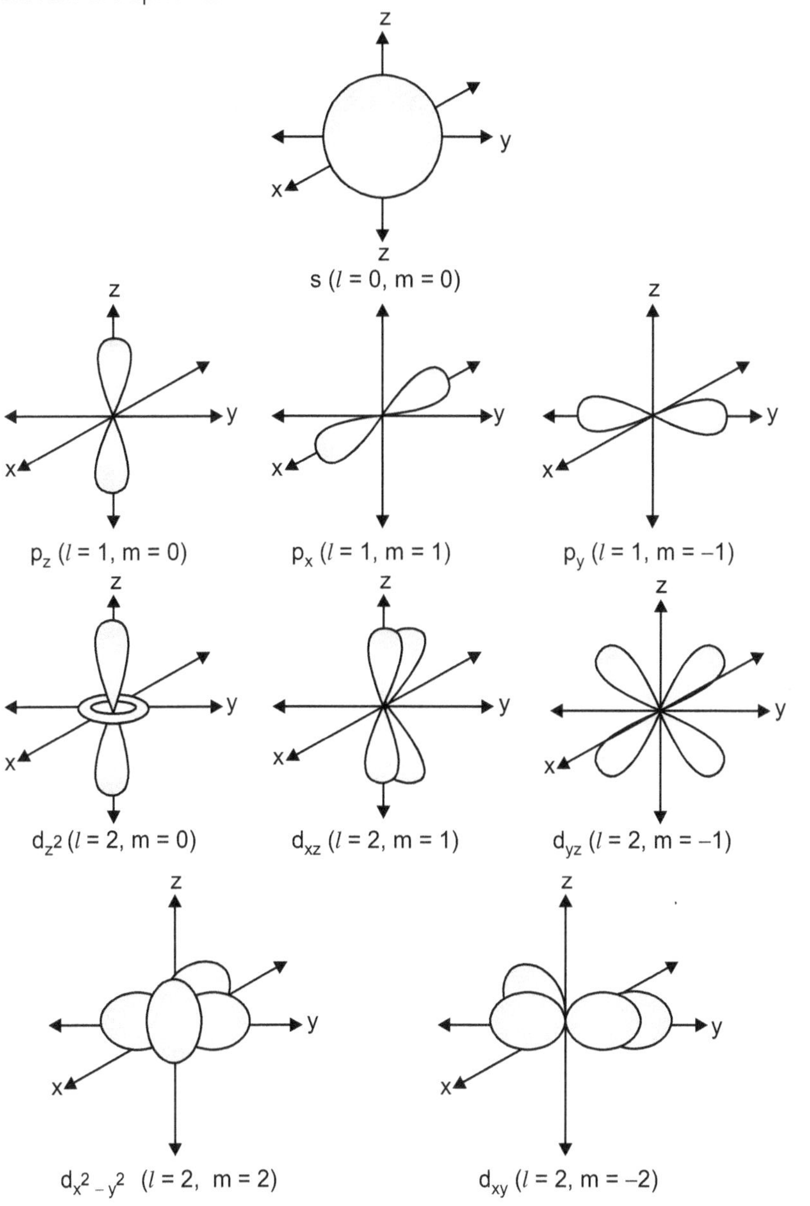

Fig. 1.16: Plots of the angular parts of the hydrogen atom wave function

The p orbitals consist of two portions which together resemble a dumbbell. Although equivalent, they differ from each other in their orientations along the axis. The d orbitals differ both in shape and space orientation.

1.14 RADIAL AND ANGULAR NODES AND THEIR SIGNIFICANCE

Representation of orbitals with higher values of n (principal quantum number) and *l* (azimuthal quantum number) becomes very complicated.

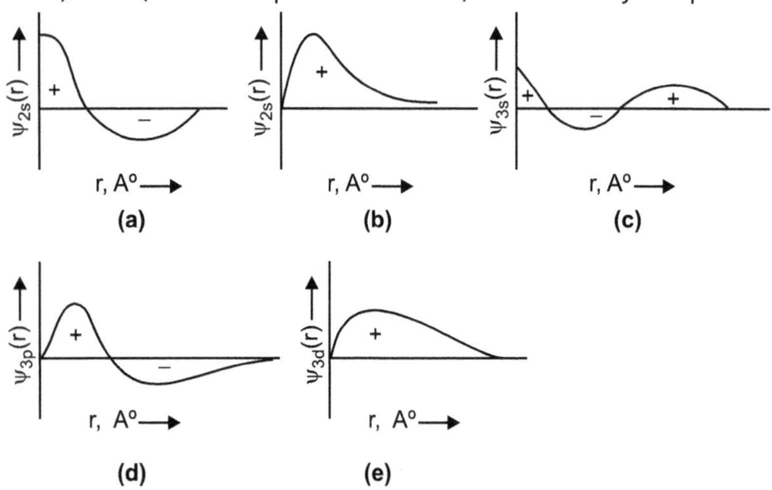

Fig. 1.17: The radial dependence of the wave functions for the various excited states

The various plots are shown in Fig. 1.17. We know that $\psi_{(1s)}$ is positive everywhere, the wave functions of other orbitals have some regions where they are positive and some others where they are negative as can be seen from the radial and angular plots in Fig. 1.17. The signs have only relative significance. ψ^2 is of course, always positive irrespective of the sign of ψ.

The positive and negative regions are separated either by a point or a line or a surface where the probability is zero. These are called nodal points, nodal line or nodal surface respectively. Some one may ask as to how an electron goes from one place of finite probability to another place of infinite probability and yet have zero probability in between, he must remember that the trajectory or path of an electron has no meaning within the quantum framework.

All the orbitals $l \neq 0$ have angular dependence. Therefore, p and d and other higher angular momentum orbitals are not spherically symmetrical. Refer Fig. 1.18. The angular dependence of the three p orbitals is such that they have the same shape, but are oriented along the three mutually perpendicular directions. The orbitals are named as p_x, p_y and p_z orbitals. The radial distribution function of the 2s orbital shows two maxima - a smaller one near the nucleus and the bigger one at a longer distance.

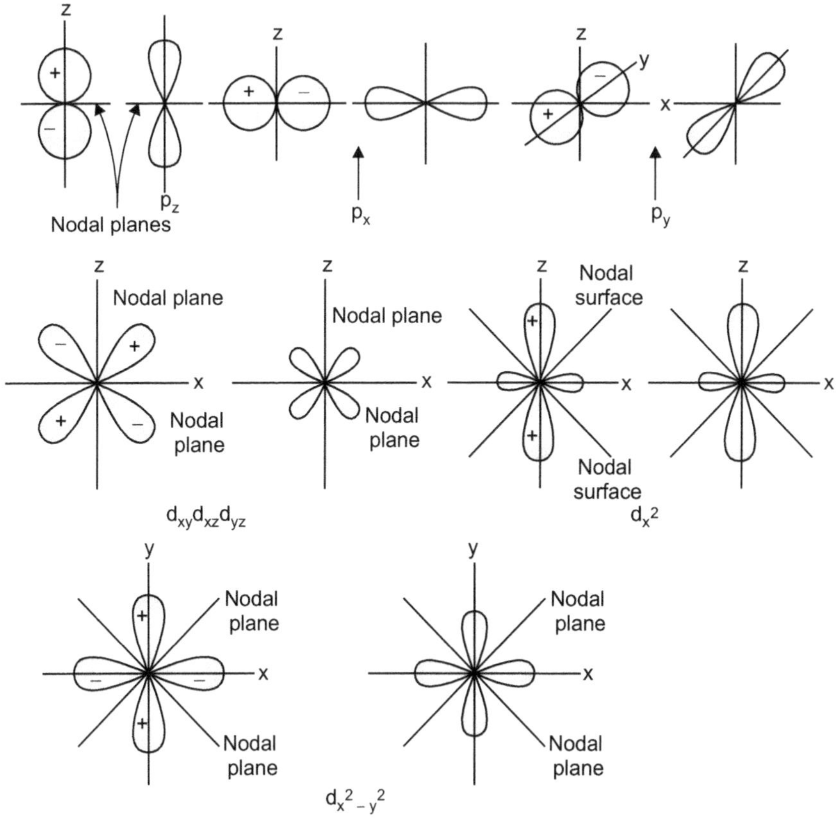

Fig. 1.18: The angular dependence of wave functions and of the probability functions for various orbitals

The radial distribution of the three 2p-orbitals is identical. It shows only one maximum which is at a greater distance than the 1s maximum. This means that 2p electron is further away from a 1s electron and has a much smaller probability as compared to 2s electron to be found near the nucleus.

1.15 RADIAL DISTRIBUTION FUNCTIONS AND THE CONCEPT OF THE MOST PROBABLE DISTANCE WITH SPECIAL REFERENCE TO 1s AND 2s ATOMIC ORBITALS

With the help of quantum mechanical wave function ψ, the radial distribution function for 1s orbital can be explained with the help of a plot of ψ_{1s} (r) versus distance (r) from the nucleus, for fixed θ and ϕ. This type of variation is known as radial dependence. Refer Fig. 1.19 (a). This plot shows that how $\psi_{(1s)}$ varies with distance from the nucleus along a fixed direction on one side of the nucleus.

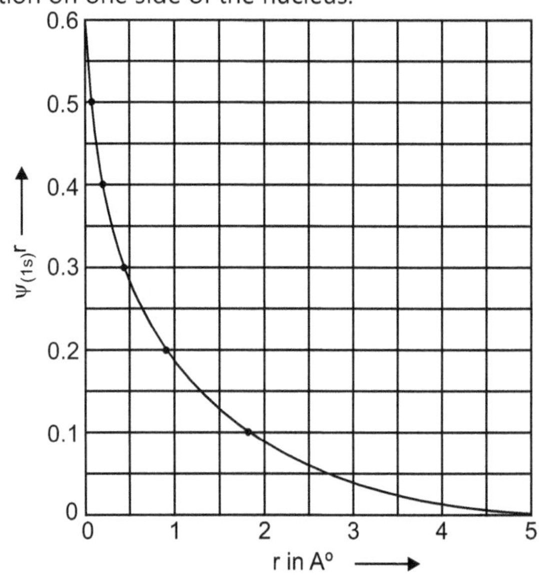

Fig. 1.19 (a): The dependence of the 1s wave function $\psi_{(1s)}$ plotted as a function of distance (r) from the nucleus

The radial dependence of $\psi_{(1s)}$ on both sides of the nucleus is shown in Fig. 1.19 (b). The radial plots do not have any direct physical significance, but are useful in molecular structure because atomic wave functions are needed to build the molecular wave functions.

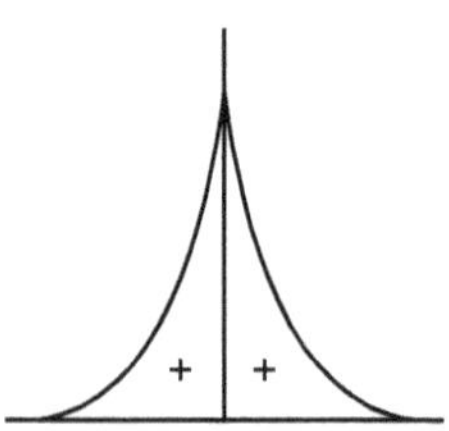

Fig. 1.19 (b): Dependence of ψ_{1s} on r on both sides of the nucleus in a particular direction

Fig. 1.19 (c) is the radial dependence of $\psi^2_{(1s)}$ along a definite direction on one side of the nucleus. Since $\psi^2_{(1s)}$ measures probability, this graph shows that the probability of finding the electron decreases rapidly as the distance from the nucleus increases. The probability never strictly becomes zero, no matter how large the distance is, but it does become very small beyond 1.5 A.

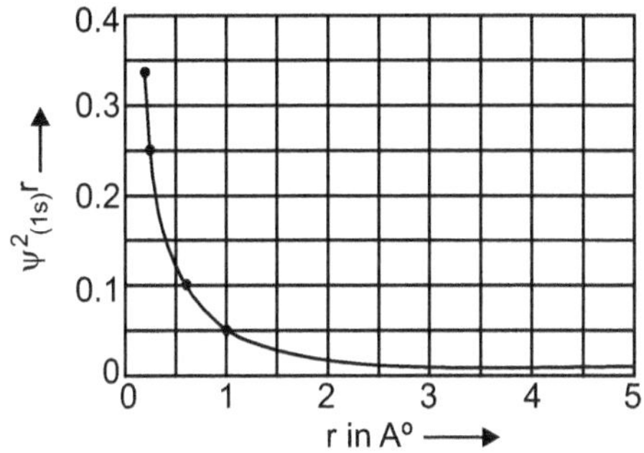

Fig. 1.19: (c) The probability of finding the electron as a function of r in a particular direction

In Fig. 1.19 (d), each point on the curve gives the probability at a particular distance for a definite direction. Sometimes, it is more useful to know the total probability at a particular distance irrespective of direction. This is obtained by adding the probabilities in all directions for that particular distance. This type of function is known as radial distribution function for the 1s orbital as shown in Fig. 1.19 (d). Mathematically, Fig. 1.19 (d) is obtained by multiplying the ordinate of Fig. 1.19 (c) by $4\pi r^2$ which is the surface area of a sphere of radius r.

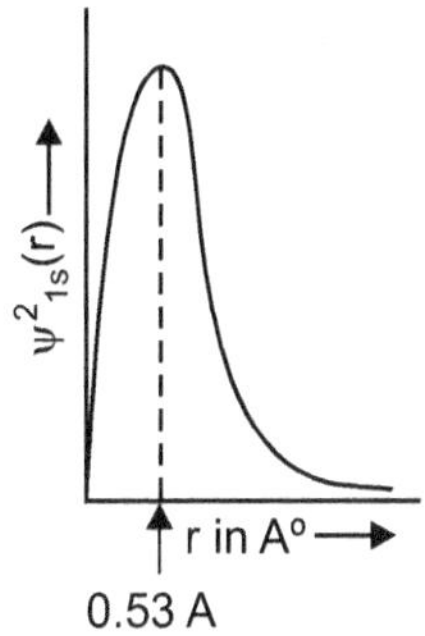

Fig. 1.19: (d) The radial distribution function for the electron in the ground state

The maximum in the radial distribution function indicates the most probable location of the electron. For the ground state, the maximum occurs at 0.53 Å, the same distance as the radius of the first Bohr's orbit. Although in the quantum picture of the hydrogen atom, the electron can be at different distances with different probabilities, the maximum probability is still at the same distance as in Bohr's model.

Radial distribution functions are most useful in energy considerations since electron-nucleus potential energy (e^2/r) depends on distance only and is independent of direction. Radial distribution functions therefore help us to understand the relative energies of orbitals for multielectron atoms.

The radial distribution function of the 2s orbital (Fig. 1.20) shows two maxima, a smaller one near the nucleus and bigger one at a larger distance.

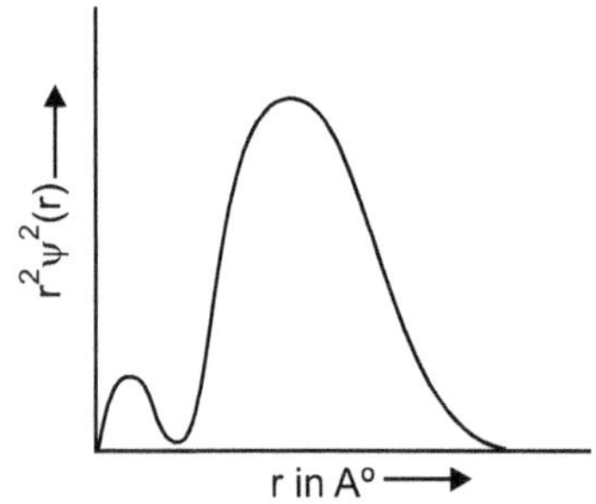

Fig. 1.20: Radial distribution function for 2s orbital

Comparing the location of the maximum in the 2s orbital with the maximum in the 1s orbital, we conclude that an electron in the 2s orbital has greater probability to stay further away than an electron in the 1s orbital. The distance at which the largest peak occurs increases with the principal quantum number. This means that the probability of the electron being at larger distances increases as the energy of the orbital increases. The radial distribution function of the three 2p-orbitals is identical. It shows only one maximum which is at a greater distance than the 1s maximum. This means that 2p electron is further away from a 1s electron and had a much smaller probability as compared to a 2s electron to be found near the nucleus. For this reason, we say that a 2s orbital penetrates closer to the nucleus than a 2p orbital. In general, higher the angular momentum, lesser is the penetration close to the nucleus.

1.16 SIGNIFICANCE OF QUANTUM NUMBERS, ORBITAL ANGULAR MOMENTUM AND QUANTUM NUMBERS m_l AND m_s

We know that, Niel's Bohr modified the Rutherford's atomic model on the basis of some assumptions. He considered the quantum mechanical condition that the angular momentum of an electrons is an integral multiple of $h/2\pi$. Thus, for the first time he has taken the help of quantum condition and introduced a quantum number named by him as principal quantum number (n) to describe electron orbit and energy of an electron. The fine structure of spectral lines indicates that, for any given quantum number 'n' there can be several sublevels of slightly different energies. Each sublevel may have its own quantum number. Sommerfield introduced the azimuthal quantum number to explain movement of electrons in an elliptical orbit. The spectroscopic study has shown that to characterize each electron in the atom, two more quantum numbers are needed. Thus, in order to designate electron in an atom, four quantum numbers are needed and these are: (i) Principal quantum number 'n', (ii) Azimuthal or Subsidiary quantum number 'l', (iii) Magnetic quantum number 'm' and (iv) Spin quantum number 's'.

(i) Principal quantum number 'n':

This number indicates the principal or major level to which the electrons belongs. It was introduced by Niel's Bohr. The principal quantum number have all integral values as n = 1, 2, 3, 4 ...... etc. Thus, when n = 1, is called first orbit, n = 2, second orbit and so on. Thus, it represents average size of electron orbit and the size of orbit increases with in value of 'n'. It can also be designated by alphabets like n = 1, K shell, n = 2, L shell, n = 3, M shell, n = 4, N shell and so on. This quantum number is also used to designate the energy levels. n = 1 first energy level, n = 2 second energy level and so on.

(ii) Azimuthal quantum number (l):

This quantum number was introduced by Sommerfield and it depends on principal quantum number n. For any given value of 'n', the azimuthal quantum number have all integral values from 0 to (n – 1). Each value refers to an energy sublevel. For example, if n = 1, l have only one value i.e. l = 0, if n = 2 then l have two values i.e. l = 0, 1, if n = 3, l = 3 i.e. 0, 1, 2 and if n = 4, 'l' have four values i.e. l = 0, 1, 2 and 3.

Thus, the sublevel given by 'l' is called an orbit. If l = 0, it is called as "s orbital". Similarly for l = 1, p orbital, l = 2, d orbital, l = 3, f orbital are possible. The number also gives the knowledge of shape of electronic orbital. Thus, 's orbital' is spherical, 'p orbitals' are dumb bell shaped.

(iii) Magnetic quantum number (m):

When the atomic spectra is studied by keeping the source in the electric or magnetic field then it is observed that the signal splits into number of lines or fine structure. Thus, due to the application of electric field the splitting takes place and is called stark effect. While by application of magnetic field when splitting is observed is known as Zeeman effect. Refer Fig. 1.21. The electron can have different possible orientations in the magnetic field for the same angular momentum. The splitting up of spectral lines may be due to these different possible orientations. For given value of 'l' the magnetic quantum number 'm' will have ($2l$ + 1) values. It can have all integral values between $-l$ to $+l$ including zero. Thus, if l = 0, m = 0, l = 1, then m = −1, 0, +1. If l = 2 then m = −2, −1, 0, +1, +2 and so on.

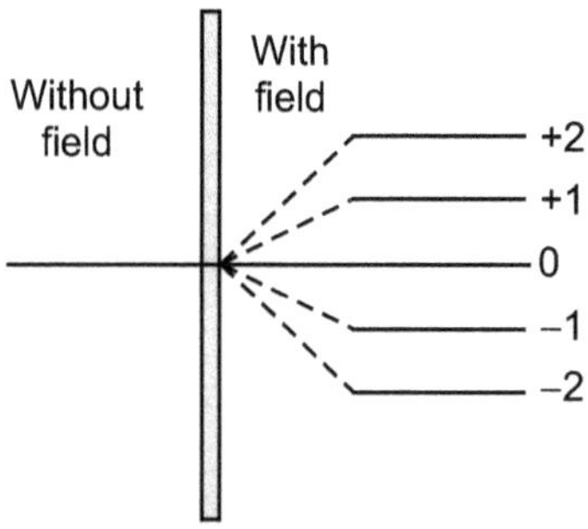

Fig. 1.21: Zeeman effect

(iv) Spin quantum number (s):

The electron is spinning about its own axis. To account for it, spin quantum number is required. For spinning motion only two directions are possible i.e. clockwise or anticlockwise. Therefore, for spin quantum number for any value of n, l and m, it has only two permissible values i.e. $+\frac{1}{2}$ or $-\frac{1}{2}$.

Thus, with the help of these four quantum numbers, the electrons in an atom may be located. We can summarise for some given values of n, the values of l, m and s in the following Table 1.2.

Table 1.2: Quantum numbers

Sr. No.		1	2	3			
1	n	K-shell	L-shell	M-shell			
2	l	0(s)	0(s)	1(p)	0(s)	1(p)	2(d)
3	m	0	0	−1, 0, +1	0	−1, 0, +1	−2, −1, 0, +1, +2
4	s	$\pm\frac{1}{2}$	$\pm\frac{1}{2}$	$\pm\frac{1}{2}, \pm\frac{1}{2}, \pm\frac{1}{2}$	$\pm\frac{1}{2}$	$\pm\frac{1}{2}, \pm\frac{1}{2}, \pm\frac{1}{2}$	$\pm\frac{1}{2}, \pm\frac{1}{2}, \pm\frac{1}{2}, \pm\frac{1}{2}, \pm\frac{1}{2}$

1.17 SHAPES OF s, p AND d ATOMIC ORBITALS

According to the concept of wave mechanics, the negative charge of electron is spread round the nucleus. This is usually referred to as 'cloud of negative charge'. The charge density can vary and it is maximum in particular region. These regions of higher probability of charge clouds are called orbitals. It can also be considered as a region at which there is higher probability that electron will be found with certain fixed amount of energy. In other words, the orbital represents most probable area where electron stays for large time when it is in constant motion.

The principal quantum number determines the size of the orbital, azimuthal quantum number governs the shape and magnetic quantum number describes the orientations. These quantum numbers should be taken into consideration to discuss the characteristics of orbitals.

s-orbital: When l = 0, only value possible for m is zero. It means for each value of principal quantum number n, there is only one orientation for l = 0. Thus, 's' orbital corresponds to spherical shape with atomic nucleus at its centre. For every value of n, there is one spherically symmetrical orbit. Its radius depends on the value of n.

p-orbitals: When l = 1, the magnetic quantum number will have three values, +1, 0, −1. So there are three orientations for p orbital. These orbitals are dumbbell shaped. Orbitals of sublevel have similar shape but they are oriented in three different directions along x, y and z axis of usual co-ordinate system. These orbitals are designated as p_x, p_y and p_z orbitals.

d-orbitals: When the principal quantum number n = 3, the azimuthal quantum number l will have three values i.e. 0, (s-orbital) 1, (p orbital) and 2 (d orbital). The d orbital have permissible orientations given by magnetic quantum number m, and they are −2, −1, 0, +1, +2. Thus, d orbitals are double dumbbell shaped. All the s, p and d orbitals are shown diagrammatically in Fig. 1.22.

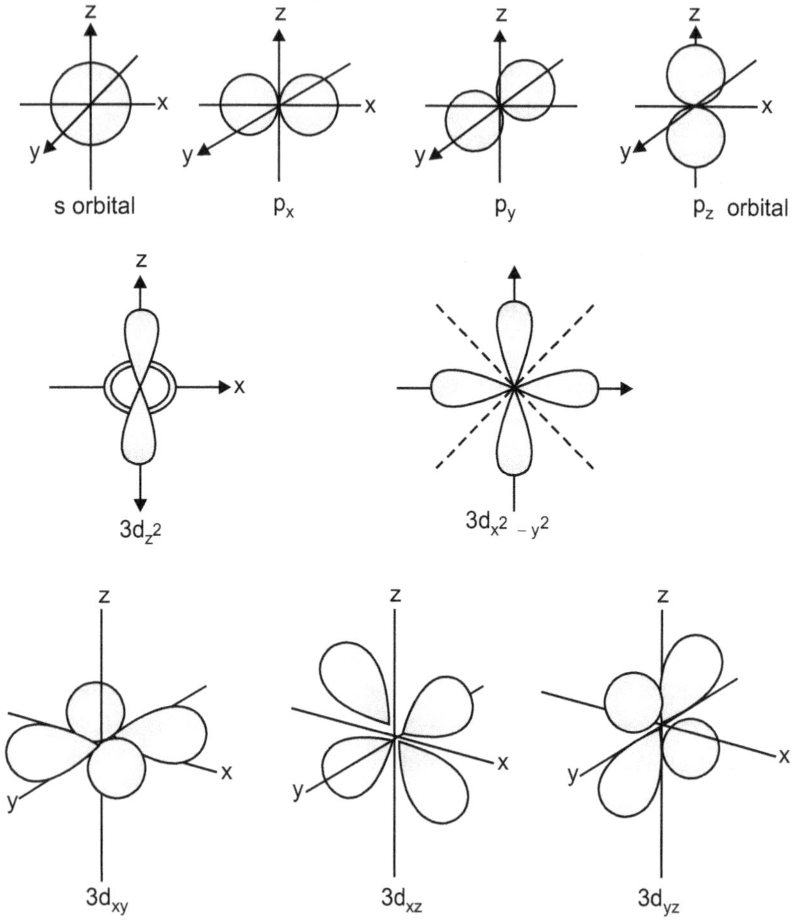

Fig. 1.22: Shapes of 's', 'p' and 'd' orbitals

1.18 DISCOVERY OF SPIN, SPIN QUANTUM NUMBER (s) AND MAGNETIC SPIN QUANTUM NUMBER (m_s)

The scientist Stern-Gerlach perform this classic experiment, carried out in 1921 which was designed to distinguish between quantum mechanical model of the atom proposed by Niel's Bohr and classical planetary models. This experimental set up (Fig. 1.23) consists of a silver beam generated by an oven in a vacuum chamber was collimated by two narrow slits 0.03 mm width. The beam is passed through an inhomogeneous magnet 3.5 cm in length and impinged on a glass plate.

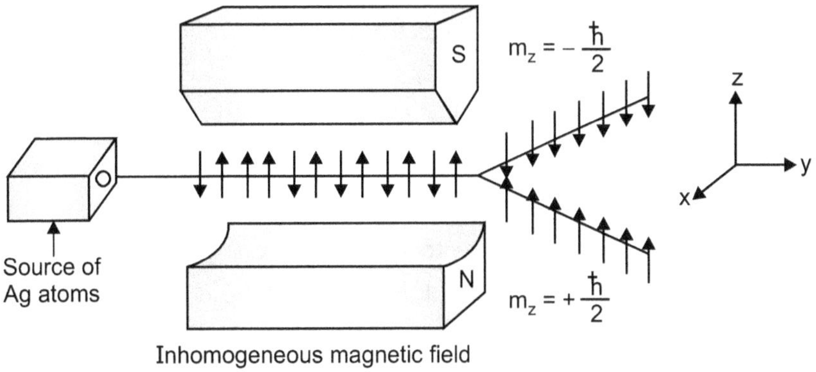

Fig. 1.23: Experimental set up of Stern-Gerlach experiment.

The inhomogeneous magnetic field separates beam into two and only two components

After about an hour of operation the plate was removed and examined visually. Only about one atomic layer of Ag was deposited on the plate in this time, making the detection of the spatial distribution of the silver atoms very difficult. Therefore, they are used as a simple technique to visualise the deposited Ag atom by passing smoke of cheap cigars with high sulphur content on the beam of Ag atoms. The sulphur containing smoke is reacted with Ag atoms, producing Ag_2S which was clearly visible, under a microscope, even though the amount deposited was less than 10^{-7} mol. They successfully completed the experiment and obtained the results without magnetic field and with magnetic field. Refer Fig. 1.24. The splitting of the beam into two distinct components is clearly visible.

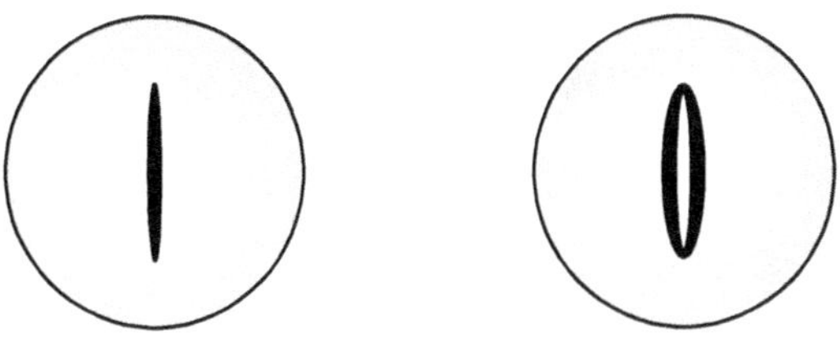

(a) Without magnetic field **(b) With magnetic field**

Fig. 1.24

Although the result did not confirm the classical model of the atom, several years later, researchers discovered that the electron has an intrinsic angular moment (spin). This angular moment and not a magnetic moment produced by electrons orbiting around the nucleus is the basis for the deflection observed.

Thus, Stern and Gerlach observed two bands which may be due to atoms orbital angular momentum (which arises from the motion of electron around the atomic nucleus). This was their opinion at that time. But afterwards researchers come to know that the angular momentum is of the motion of electron about its own axis. This internal angular momentum of the electron is called its spin. In order to distinguish this spin angular momentum from orbital angular momentum, we use quantum numbers 's' and m_s. The magnitude of the spin angular momentum is $\{s(s + 1)\}^{1/2} \frac{h}{2\pi}$ and the component $\frac{m_s h}{2\pi}$ is restricted to the (2s + 1) values given by

$$m_s = s, s - 1, s - 2, \dots -s$$

For an electron, only one value of s is allowed and that $s = \frac{1}{2}$ corresponding to an angular momentum. The spin angular momentum is an intrinsic property of the electron, like its rest mass and charge, and every electron has exactly the same value. The spin may lie in any of 2s + 1 = 2 different orientations. One orientation corresponds to $m_s = +\frac{1}{2}$ and the other corresponds to $m_s = -\frac{1}{2}$.

SOLVED EXAMPLES

Example 1.2: *Calculate the wavelength of tennis ball (mass 140 gm) moving with a velocity of 250 cm/s.*

Solution: (i) Mass of tennis ball = 140 gm

(ii) Velocity of the ball = 250 cm/s

The momentum of the ball = Mass × Velocity

$$mv = 140 \times 250$$

$$\text{Momentum (mv)} = 35000 = 3.5 \times 10^4 \text{ g cm/s}$$

Therefore, from the de-Broglie relation, the wavelength, λ, should be

$$\lambda = \frac{h}{p} \qquad p = mv = \text{Momentum}, \quad h = 6.63 \times 10^{-27} \text{ erg sec.}$$

$$\lambda = \frac{6.63 \times 10^{-27}}{3.5 \times 10^{4}} = 1.89 \times 10^{-31} \text{ cm}$$

$$\lambda = \mathbf{1.89 \times 10^{-31}\ cm}$$ **... Ans.**

Example 1.3: *Calculate the wavelength of an electron of mass 9.1×10^{-28} gm moving with a velocity of 10^8 cm/s.*

Solution: (i) Mass of electron = 9.1×10^{-28} gm

(ii) Velocity of electron = 10^8 cm/s

The momentum of electron is calculated by using the equation

$$\text{Momentum} = m \cdot v$$

$$\text{Momentum (p)} = 9.1 \times 10^{-28} \times 10^{8} = 9.1 \times 10^{-20} \text{ g cm/s}$$

By using de-Broglie's wave equation

$$\lambda = \frac{h}{p} \qquad h = 6.63 \times 10^{-27} \text{ erg sec}, \quad p = 9.1 \times 10^{-20} \text{ g cm/s}$$

$$\lambda = \frac{6.63 \times 10^{-27}}{9.1 \times 10^{-20}} = 7.26 \times 10^{-8} \text{ cm}$$

$$\lambda = \mathbf{7.26 \times 10^{-8}\ cm}$$ **... Ans.**

Example 1.4: *If an electron is to be located within 0.1 A (1×10^{-9} cm), what will be the approximate uncertainty in its velocity ?*

Solution: According to uncertainty principle,

$$\Delta p \times \Delta x = \Delta p \times \left(\frac{h}{\Delta p}\right) = h$$

The approximate uncertainty in velocity (Δv) will be given by

$$\Delta v = \frac{\Delta p}{m} = \frac{h}{m\Delta x} = \frac{6.63 \times 10^{-27}}{9.1 \times 10^{-28} \times 10^{-9}}$$

$$\Delta v = \mathbf{7.29 \times 10^{9}\ cm/s}$$ **... Ans.**

Example 1.5: *If a cricket bass (mass 140 g) is to be located within 0.1 A, what will be the approximate uncertainty in its velocity ?*

Solution: The equation for uncertainty in velocity is

$$\Delta v = \frac{\Delta p}{m} = \frac{h}{m\Delta x}$$

$h = 6.63 \times 10^{-27}$ erg sec
$m = 140$ g
$\Delta x = 0.1 \times 10^{-8}$ cm

$$\Delta v = \frac{6.63 \times 10^{-27}}{140 \times 1 \times 10^{-9}}$$

$$\Delta v = \mathbf{4.74 \times 10^{-20}\ cm/s}$$ ... **Ans.**

Example 1.6: *Calculate the radius of the first Bohr obit.*

Solution: The equation to obtain r given by Bohr is

$$r_n = \frac{n^2h^2}{4\pi^2 mZe^2}$$

$$r_n = \frac{1 \times (6.625 \times 10^{-27})^2}{4 \times (3.142)^2 \times 9.1 \times 10^{-28} \times (4.80 \times 10^{-10})^2}$$

$$r_n = 0.529 \times 10^{-8} \text{ cm}$$

$$= \mathbf{0.529\ A°}$$ ... **Ans.**

First Bohr orbit means
$n = 1$, $h = 6.625 \times 10^{-27}$ erg sec,
$\pi = 3.142$, $Z = 1$,
$m = 9.1 \times 10^{-28}$ gm,
$e = 4.80 \times 10^{-10}$ esu

Example 1.7: *What will be the wavelength of radiation emitted if an electron transmission takes place from third energy level to second energy level in hydrogen atom. (R_H = log 677 cm^{-1}).*

Solution: $$\bar{v} = \frac{1}{\lambda} = R_H\left(\frac{1}{n_1^2} - \frac{1}{n_2^2}\right)$$

Let $n_1 = 2$ and $n_2 = 3$

$$\therefore \quad \frac{1}{\lambda} = \log 677 \left(\frac{1}{(2)^2} - \frac{1}{(3)^2}\right)$$

$$\frac{1}{\lambda} = \log 677 \times \frac{5}{36}$$

$$\lambda = \frac{36}{\log 677 \times 5}$$

$$\lambda = 6565 \times 10^{-8}$$

$$\lambda = \mathbf{6565\ A°}$$ ... **Ans.**

Example 1.8: *Calculate the energy of electron in the second orbit of hydrogen. Given: m = 9.109 × 10^{-28} gm, e = 4.802 × 10^{-10} esu, h = 6.625 × 10^{-27} erg sec.*

Solution: The energy of electron in the orbit is calculated by the formula

$$E = -\frac{2\pi^2 mZ^2e^4}{n^2h^2}$$

For hydrogen, Z = 1 and for second orbit, n = 2.

$$E = -\frac{2 \times (3.142)^2 \times (9.109 \times 10^{-28}) \times (4.802 \times 10^{-10})^4}{(2)^2 \times (6.625 \times 10^{-27})^2}$$

$$E = -5.444 \times 10^{-11} \text{ ergs.}$$

This is also expressed in electron volts.

$$1 \text{ eV} = 1.602 \times 10^{-12} \text{ ergs}$$

$$E = \frac{-5.444 \times 10^{-11}}{1.602 \times 10^{-12}} = \mathbf{-3.399 \text{ eV}}$$ **... Ans.**

PRACTICE QUESTIONS

(A) Multiple Choice Questions :

1. The combined feature that distinguishes quantum and classical mechanics is ....
   (a) uncertainty principle (b) wave particle duality
   (c) photoelectric effect (d) black body radiation
2. The model of an atom was putforth for the first time by ....
   (a) Rutherford (b) Niel's Bohr
   (c) Sommerfield (d) J. J. Thomson
3. Total amount of energy 'E' radiated by a black body per unit area and time is given by ......
   (a) Stefan-Boltzmann fourth power law
   (b) Wien's displacement law
   (c) Rayleigh-Jean's law
   (d) Einstein's law
4. The de-Broglie's wave particle duality equation first supported by experimental proof was ......
   (a) George Thomson (b) Germer and Davisson
   (c) Niel's Bohr (d) Max Planck
5. The hydrogen atomic spectra is well explained by ......
   (a) J. J. Thomson model (b) Rutherford's model
   (c) Bohr's model (d) Sommerfield's model
6. The principal quantum number (n) was introduced by ....
   (a) Max Planck (b) Einstein
   (c) Rayleigh (d) Niel's Bohr

7. Uncertainty principle was given by ......
   (a) Einstein (b) Max Planck
   (c) Werner Heisenberg (d) Newton
8. The operator for energy is called ......
   (a) Laplacian operator (b) Hamiltonian operator
   (c) Schrodinger's operator (d) none of these
9. The wave function for an electron in an atom is called as ....
   (a) atomic orbital (b) orbital
   (c) angular part (d) probability
10. The positive and negative regions are separated either by a point or a line or a surface where the probability is zero is known as ....
   (a) a plane (b) nodal point
   (c) trajectory (d) nodal angle
11. The maximum in the radial distribution function indicates the most probable location of the ....
   (a) atom (b) neutron
   (c) emission (d) proton
12. The principal quantum number (n) gives the information about ....
   (a) shape of orbit (b) size of orbit
   (c) length of orbit (d) geometry of orbit
13. The subsidiary quantum number (l) is useful for giving the information about the ....
   (a) size (b) shape
   (c) length (d) depth
14. For spinning of electron number, possible orientations are ....
   (a) one (b) two
   (c) three (d) four
15. The shape of 1s orbital is ....
   (a) spherical (b) rectangular
   (c) triangular (d) none of these

ANSWERS

1. (b)	2. (d)	3. (a)	4. (b)	5. (c)	6. (d)	7. (c)	8. (b)
9. (a)	10. (b)	11. (c)	12. (b)	13. (b)	14. (b)	15. (a)	

(B) State True or False:

1. The biological science is increasingly focused on the molecular level.
2. A black body is not only a perfect absorber of radient energy, but also a perfect radiator.
3. No electron is emitted unless the frequency ν is below a threshold frequency ν_0 even for high light intensities.
4. The de-Broglie's relation shows that the wavelength is directly proportional to the momentum of a particle.
5. The diffraction phenomena is due to wave nature of the matter.
6. Atomic spectra is always a line spectra.
7. According to Niel's Bohr, angular momentum is an integral multiple of h.
8. The product of uncertainties in momentum and uncertainties in position cannot be less than a quantity of the order of $h/4\pi$.
9. The Schrodinger's equation must be made as wiggly in regions where V is low.
10. The square of the wave function, is the significant quantity, as it gives the probability for finding the particle at a given point in space and time.
11. The wave function for an electron in an atom is called atomic orbital.
12. The ψ^2 is always negative irrespective of the sign of ψ.
13. The maximum in the radial distribution function indicates the most probable location of the electron.
14. To designate the electron in an atom, the knowledge of four quantum numbers is necessary.
15. The subsidiary quantum number does not give the orientation of orbitals.

Ans.: 1. True, 2. True, 3. False, 4. False, 5. True, 6. True, 7. False, 8. False 9. False, 10. True, 11. True, 12. False, 13. True, 14. True, 15. False.

(C) Short and Long Answer Questions:

1. Discuss the history of quantum mechanics.
2. The discrepancy between experiments and classical theory for black body radiation was at high and not at low frequencies. Explain.
3. What is ultraviolet catastrophe?
4. Write a brief account on black body radiation.
5. Discuss the photoelectric effect.
6. Explain the concept of wave particle duality. What are de-Broglie's waves ?

7. State and explain uncertainty principle.
8. State and explain postulates of Bohr's theory.
9. Derive an expression for radius and energy of an hydrogen atom with the help of Bohr's theory.
10. Give the limitations of Bohr's theory.
11. Derive the time independent Schrodinger's equation.
12. Discuss significances of ψ and ψ^2.
13. How the Schrodinger's equation is applied to hydrogen atom ?
14. Give some general conclusions from the mathematical treatment to hydrogen atom.
15. Explain the radial and angular parts of hydrogenic wave functions with graphical representation.
16. Write a brief account on radial and angular nodes.
17. What is radial distribution function ? Explain with reference to 1s and 2s orbitals.
18. Discuss quantum numbers in detail.
19. Write a note on shapes of s, p and d atomic orbitals.
20. Explain the experiment carried out by Stern-Gerlach.

UNSOLVED PROBLEMS

1. The speed of an electron is 1.2×10^{10} cm/sec. Calculate the de-Broglie wavelength of the electron. (**Ans.** 6.0×10^{-10} cm)
2. Wavelength of light in yellow region is 5890 Å. Calculate the energy in one photon of yellow light. (**Ans.** 3.37×10^{-12} ergs)
3. Calculate the radius of second Bohr orbit. (**Ans.** 2.116 Å)
4. Calculate the velocity of electron in the first Bohr orbit. (**Ans.** 2.188×10^{8} cm/s)
5. Calculate the energy of an electron in the first Bohr orbit. (**Ans.** −3.402 eV)
6. Calculate the wavelength of cricket ball (mass 150 g) moving with velocity 200 cm/s. (**Ans.** 2×10^{-31} cm)
7. If an electron is to be located within 0.2 Å (2×10^{-9} cm), what will be the approximate uncertainty in its velocity ? (**Ans.** 3.7×10^{9} cm/s)

***

Chapter **2**...

Periodic Table and Periodicity of Elements

Contents ...

2.1 INTRODUCTION

Every atom consists of a central nucleus surrounded by one or more orbital electrons. The nucleus contains protons and neutrons (exception, nucleus of hydrogen contains only one proton). The protons and neutrons are responsible for the mass of atom. Both protons and neutrons are particles of unit mass, but proton has one positive charge while neutron is electrically neutral. Thus, nucleus of an atom is always positively charged. This positive charge on the nucleus is exactly balanced by an equal number of orbital electrons, each of which has one negative charge.

The study of distribution of electrons around the nucleus is utmost important because it is this property of element that decides its chemistry. **'The arrangement of electrons in various shells, subshells and orbitals of an atom in its ground state is called its electronic configuration'**. The knowledge of electronic configuration is very useful in the study of chemistry. The properties of element depend upon the number of electrons and their arrangement around the nucleus, that is, on electronic configuration of an atom.

According to Bohr, electrons revolve around the nucleus in a specified path called **orbit**. Electron in each such orbit has a fixed energy. Hence, various orbits were called energy levels. Starting from level close to the nucleus these are numbered as 1, 2, 3, ... etc. The energy levels are also called shells. The shells with numbers 1, 2, 3, 4 etc. are called as **K, L, M, N shells**. Bohr further stated that the **maximum number of electrons in an orbit (shell) or energy level is $2n^2$, where 'n' is principal quantum number** of that orbit. Following Table 2.1 shows various shells, their quantum numbers and maximum number of electrons in each shell.

Table 2.1 : Various shells and maximum number of electrons

Shell	Principal quantum number	Orbitals available	Maximum number of electrons present
K	n = 1	1s	2
L	n = 2	2s, 2p	8
M	n = 3	3s, 3p, 3d	18
N	n = 4	4s, 4p, 4d, 4f	32

Each energy level (shell) contains sublevels, that is, orbitals such as s, p, d, f, etc. **The total number of orbitals is given by n^2 where** 'n' is the principal quantum number.

2.2 QUANTUM NUMBERS

In quantum mechanics, three quantum numbers are used to describe the distribution of electrons in an atom. These numbers are derived from the mathematical solution of the Schrodinger's equation for the hydrogen atom. These quantum numbers are, (i) the principal quantum number (n), (ii) the angular quantum number (l) and (iii) the magnetic quantum number (m_l). These quantum numbers are useful to describe atomic orbitals and to label the electrons that reside in an atom. A fourth quantum number, the spin quantum number (m_s), describes the behaviour of a specific electron and completes the description of electrons in atoms.

2.2.1 The Principal Quantum Number (n)

The principal quantum number can have integral values 1, 2, 3 and so on. This indicates the number of main energy level (shell) to which the electron belongs. The principal quantum number (n) also relates to the average distance of the electron from the nucleus in a particular orbital. Larger the value of n, away is the electron from the nucleus, and therefore larger is the orbital.

2.2.2 The Angular Momentum (Azimuthal) Quantum Number (l)

The angular momentum (azimuthal) quantum number 'l' describes the shape of the orbital. The value of l depends upon the value of n. For a given value of n, l has possible values from 0 to n − 1. Thus, if n = 1, the only possible value of l = 0 (n − 1 = 1 − 1 = 0), for n = 2, there are two values of l, 0 and 1. Similarly, for n = 3, l = 0, 1 and 2. Each value of l represents one subshell. The subshells corresponding to first four values of l, that is, l = 0, 1, 2, 3 are given by letters s, p, d, f, respectively.

Thus, when

n = 1	$l = 0$	One subshell (s)
n = 2	$l = 0, 1$	Two subshells (s and p)
n = 3	$l = 0, 1, 2$	Three subshells (s, p and d)
n = 4	$l = 0, 1, 2, 3$	Four subshells (s, p, d and f)

2.2.3 The Magnetic Quantum Number (m_l)

The magnetic quantum number m_l, describes the orientation of the orbital in space. Within a subshell, the value of m_l depends upon the value of azimuthal quantum number l. For a given value of l, there are $(2l + 1)$ integral values of m_l.

For example, if $l = 0$, then $m_l = 0$, for $l = 1$ there are $(2 \times 1 + 1 = 3)$ three values of m_l namely $-l, 0, +l$ that is, −1, 0, +1. For $l = 2$, there are $(2 \times 2 + 1 = 5)$ values of m_l, namely −2, −1, 0, +1, +2. The number of m_l value indicates the number of orbitals in a subshell with a particular l value.

To know clearly about these three quantum numbers, let us consider the case in which we have n = 2 and $l = 1$. From the values of n and l, we come to know that it is a 2p subshell, and in this subshell we have three 2p orbitals as third quantum number m_l has three values −1, 0, +1.

2.2.4 The Electron Spin Quantum Number (m_s)

The fourth quantum number which describes an electron spin in an atom is the electron spin quantum number which is labelled as m_s and it has value of $+\frac{1}{2}$ or $-\frac{1}{2}$. The spin can be either clockwise or anticlockwise $\left(+\frac{1}{2} \text{ or } -\frac{1}{2}\right)$.

The possible set of first three quantum numbers is shown in Fig. 2.1 below.

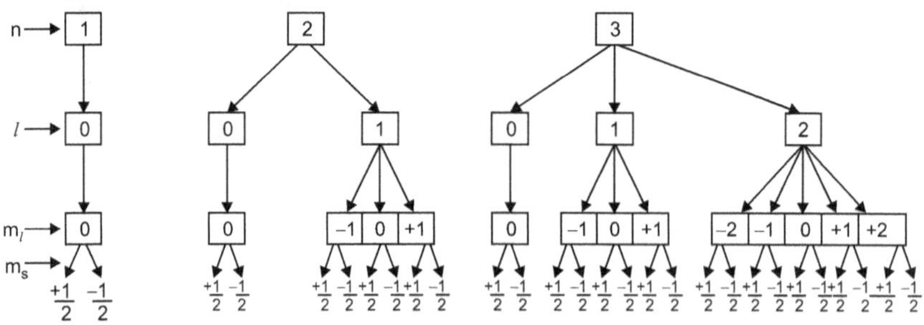

Fig. 2.1 : The possible sets of quantum numbers for n = 1, n = 2 and n = 3

2.3 SHAPES OF ATOMIC ORBITALS

The subsidiary quantum number l describes the shape of the orbital occupied by the electron. l may have values 0, 1, 2, or 3. When $l = 0$, the orbital is spherical and is called an s orbital; when $l = 1$, the orbital is dumb-bell shaped and is called a p orbital; when $l = 2$, the orbital is double dumb-bell shaped and is called a d orbital; and when $l = 3$, a more complicated f orbital is formed. The letters s, p, d and f derived from the spectroscopic terms sharp, principal, diffuse and fundamental which were used to describe the lines in the atomic spectra.

The basic question that is asked while studying the properties of orbitals is, what are the shapes of the orbitals ? Trully speaking an orbital does not have a well defined shape. It is thus difficult to say what an orbital exactly looks like.

In principle an electron can be found any where and most of the time it is close to the nucleus. The electron density decreases rapidly as the distance from the nucleus increases. Roughly there is about a 90% probability of finding the electron within the sphere of radium 100 picometers (1 picometer = 10^{-12} m) around the nucleus. Thus, we may represent the 1s orbital by drawing a boundary surface diagram that covers about 90% of the total electron density in an orbital.

2.3.1 s-orbital

The charge cloud of s-orbital is spread uniformly around the nucleus. The s-orbital is non-directional. All s-orbitals are spherical in shape but differ in size which increases with increase of value of principal quantum number 'n'. Thus, 2s orbital is larger than 1s orbital while 3s orbital is larger than 2s orbital. Following Fig. 2.2 shows shapes of s-orbitals.

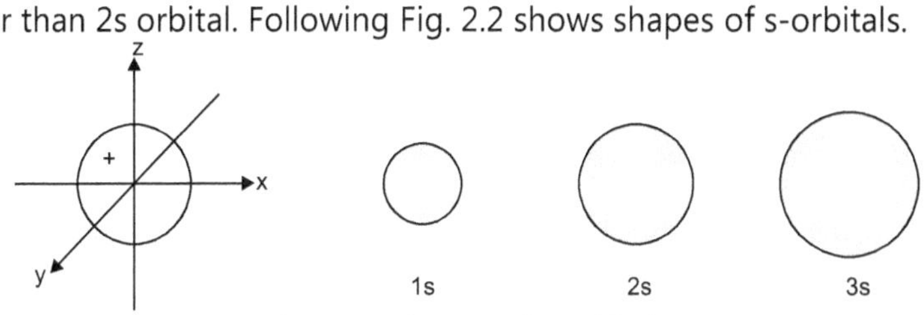

Fig. 2.2 : Shapes of s-orbitals

2.3.2 p-orbitals

The p-orbital starts with the principal quantum number n = 2. When n = 1, the angular momentum quantum number l can have only the value of zero, therefore there is only a 1s orbital. When $l = 1$, the possible m_l values are −1, 0, 1. Thus, with n = 2 and $l = 1$, we can have three 2p orbitals namely $2p_x$, $2p_y$ and $2p_z$. The letters x, y and z show the axes along which the orbitals are oriented. Following Fig. 2.3 shows boundary surface diagrams of the three 2p orbitals.

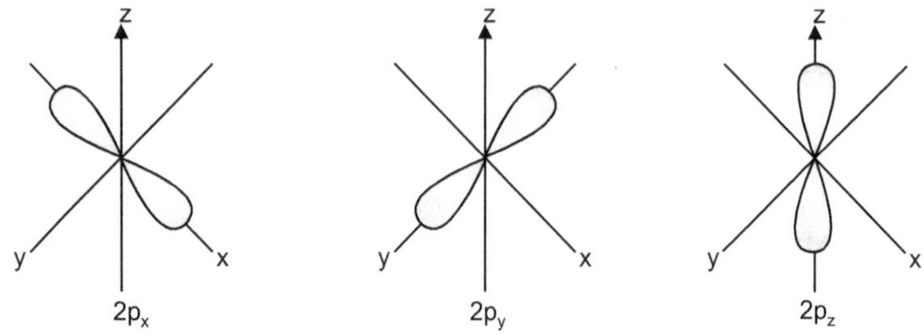

Fig. 2.3 : Shapes of p-orbitals

These three p-orbitals are identical in size, shape and energy; they differ from one another only in their orientations. The p-orbitals of higher principal quantum number have similar shapes.

Each p orbital have two lobes, the nucleus is at the centre of the p-orbital. Like s-orbital, p-orbitals increase in size from 2p to 3p to 4p orbitals and so on.

2.3.3 d-orbitals

The d-orbital starts with the principal quantum number n = 3. When n = 1 or 2, the value of l can only be 0 or 1, therefore there is only 1s, 2s and 2p orbitals. When l = 2, the possible m_l values are −2, −1, 0, +1, +2. Thus, with n = 3 and l = 2, we can have five 3d orbitals namely $3d_{xy}$, $3d_{yz}$, $3d_{xz}$, $3d_{x^2-y^2}$ and $3d_{z^2}$. Fig. 2.4 shows the shapes of d-orbitals and their orientations along x, y and z axes.

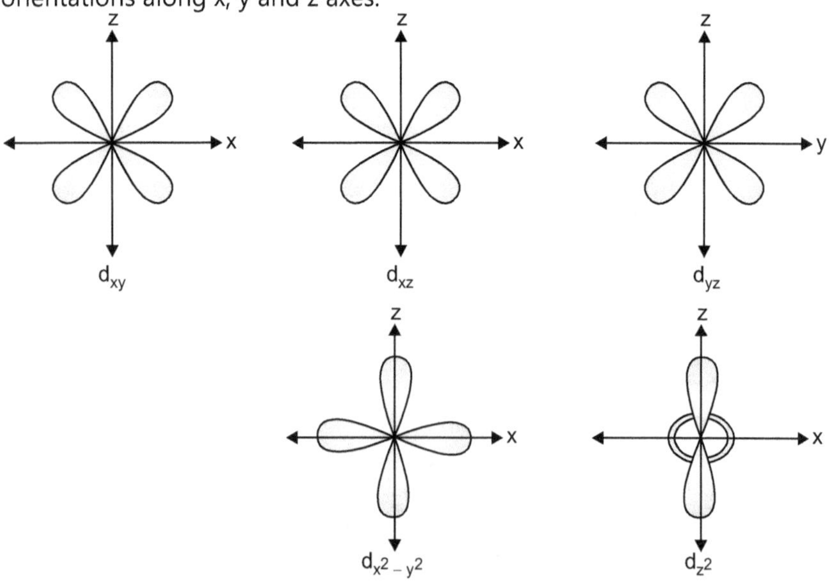

Fig. 2.4 : Shapes of five d-orbitals

As in the case of p-orbitals, the different orientations of the d-orbital correspond to the different values of m_l. All 3d orbitals in an atom are identical in energy. The d-orbitals with n value greater than 3, that is, 4d, 5d etc. have similar shapes.

2.4 THE PAULI EXCLUSION PRINCIPLE

The set of four quantum numbers n, l, m_l and m_s is used to describe a electron in any orbital and in any atom. For example, electron in 2s orbital has following four quantum numbers n = 2, l = 0, m_l = 0 and $m_s = +\frac{1}{2}$ or $-\frac{1}{2}$. Generally, the set of quantum numbers for a particular electron is given using notations as (n, l, m_l, m_s). Thus, for above case, we write $\left(2, 0, 0, +\frac{1}{2}\right)$ or $\left(2, 0, 0, -\frac{1}{2}\right)$. The value of m_s has no effect on the energy, size, shape or orientation of an orbital, but it plays important role in determining how electrons are arranged in an orbital.

There is limit to the number of electrons that can exist in a given shell. This limitation is summarized in a rule known as **Pauli exclusion principle**. This principle states that "**no two electrons in an atom can have the same set of four quantum numbers**". If two electrons in an atom have the same values of n, l and m_l then they must have different values of m_s. In other words, only two electrons may exist in the same atomic orbital, and they must have opposite spins. For example, two electrons of Helium atom can be placed in 1s orbital by using following three possibilities.

He =	↑↑	↓↓	↑↓
	$1s^2$	$1s^2$	$1s^2$
	(i)	(ii)	(iii)

Possibility numbers (i) and (ii) are ruled out by the Pauli exclusion principle. In case number (i), both electrons have similar (parallel) spin and set of quantum numbers $\left(1, 0, 0, +\frac{1}{2}\right)$, in case number (ii), both electrons have similar (anti parallel) spin and set of quantum numbers would be $\left(1, 0, 0, -\frac{1}{2}\right)$. Therefore, the only correct possibility is case number (iii), because one electron has a set of quantum numbers $\left(1, 0, 0, +\frac{1}{2}\right)$ and the other has $\left(1, 0, 0, -\frac{1}{2}\right)$. Thus, the correct configuration of He atom is,

$$He = \boxed{\uparrow\downarrow}\ 1s^2$$

Consider another example of the lithium atom, which has three electrons. First two electrons enter 1s orbital like that of He atom. The third electron cannot go into the 1s orbital because it would have the same set of four quantum numbers as one of the first two electrons $\left(1, 0, 0, +\frac{1}{2} \text{ or } 1, 0, 0, -\frac{1}{2}\right)$. Therefore, this electron enters the next higher orbital which is 2s orbital. The electron configuration of lithium is,

$$Li = \underset{1s^2}{\boxed{\uparrow\downarrow}} \quad \underset{2s^1}{\boxed{\uparrow}}$$

Thus, according to Pauli exclusion principle, any orbital can contain maximum two electrons with opposite spin. Capacity of subshell to accommodate electrons is given by equation $2 \times (2l + 1)$. Thus, for $l = 0$ (s-subshell), $2 \times (2 \times 0 + 1) = 2$. Hence, it can accommodate only two electrons. In the case of $l = 1$ (p-subshell), $2 \times (2 \times 1 + 1) = 6$, has three orbitals ($p_x$, p_y and p_z) hence, it can accommodate maximum 6 electrons. Similarly, for $l = 2$ (d-subshell), $2 \times (2 \times 2 + 1) = 10$ has five orbitals (d_{xy}, d_{yz}, d_{xz}, $d_{x^2-y^2}$ and d_{z^2}) and can accommodate maximum 10 electrons. In the case of $l = 3$ (f-subshell), $2 \times (2 \times 3 + 1) = 14$, has 7 orbitals and thus accommodate maximum 14 electrons.

2.5 BUILD-UP OF THE ELEMENTS

As we move from the 1st element hydrogen to the higher elements one by one, we see that each succeeding element would have one more proton and one extra electron than that of the preceding element. To accommodate those extra electrons into available orbitals, certain rules need to be followed. The rules are :

(i) Aufbau principle
(ii) n + l rule and
(iii) Hund's rule of maximum multiplicity.

2.5.1 Aufbau Principle

In multielectron atoms, the electrons are distributed among the orbitals of the atom according to the Aufbau (German aufbau meaning building-up) principle. It states that, **in the ground state of an atom, the orbitals of lowest energy are filled first then the orbitals of higher energy will be filled**. This means that the electrons enter the orbitals in the order of increasing energies, the lowest available orbital being filled first.

2.5.2 (n + *l*) Rule

As a working rule, a new electron enters the orbital where (n + *l*) is minimum. When (n + *l*) has the same value for two or more orbitals, the new electron enters the orbital where n is the minimum.

Table 2.2 shows (n + *l*) values for various orbitals.

Table 2.2 : n + *l* values of various orbitals

(n + *l*) value	1	2	3	4	5	6	7	8
Orbitals	1s	2s	3s 2p	4s 3p	5s 4p 3d	6s 5p 4d	7s 6p 5d 4f	8s 7p 6d 5f

The sum of n and *l* gives the energy of various possible orbitals. In neutral isolated atom the orbital with the lower value of (n + *l*) has lowest energy and being filled first. That is why 1s orbital (n = 1, *l* = 0) with n + *l* = 1 is filled first and then 2s (n = 2, *l* = 0) n + *l* = 2 is occupied by the electrons.

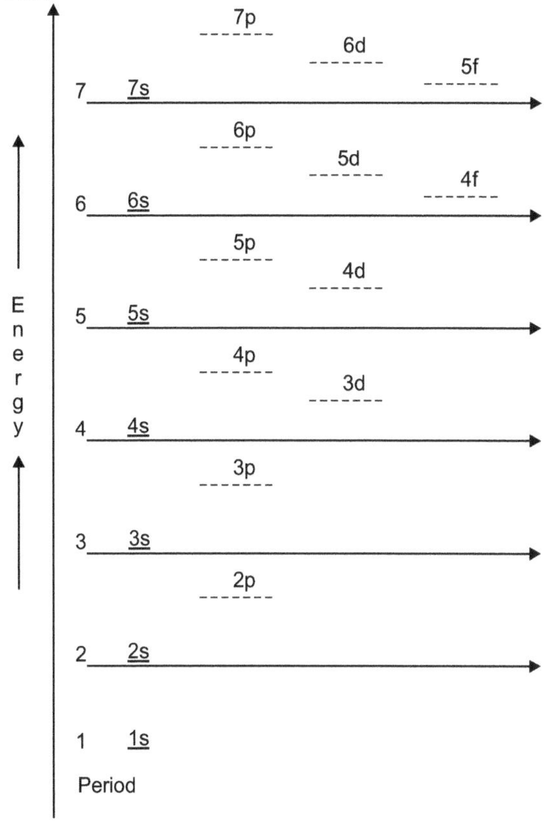

Fig. 2.5 : Comparative energies of atomic orbitals

This also explains why certain orbitals having higher value of n but lower value of l have less energy than orbitals with a lower value of n but higher value of l. For example, a 4s orbital ($n + l = 4 + 0 = 4$) has lower energy than 3d orbital ($n + l = 3 + 2 = 5$), therefore 4s orbital is occupied first and then 3d.

If the two orbitals have same ($n + l$) value then the orbital with lower n value has lower energy and filled first. For example, 3p is filled first and then 4s because 3p orbital ($n + l = 3 + 1 = 4$) and 4s orbital ($n + l = 4 + 0 = 4$) both have same values of $n + l$, but for 3p, n = 3 which is lower than 4s, n = 4.

Thus, orbitals are filled in the order of increasing value of ($n + l$). The order is, 1s, 2s, 2p, 3s, 3p, 4s, 3d, 4p, 5s, 4d, 5p, 6s, 4f, 5d, 6p, 7s, 5f, 6d, 7p .... The order of orbitals and their energies are shown in Fig. 2.5.

To understand clearly the filling order and to remember easily the order of increasing energy of orbitals, following Fig. 2.6 can be very useful.

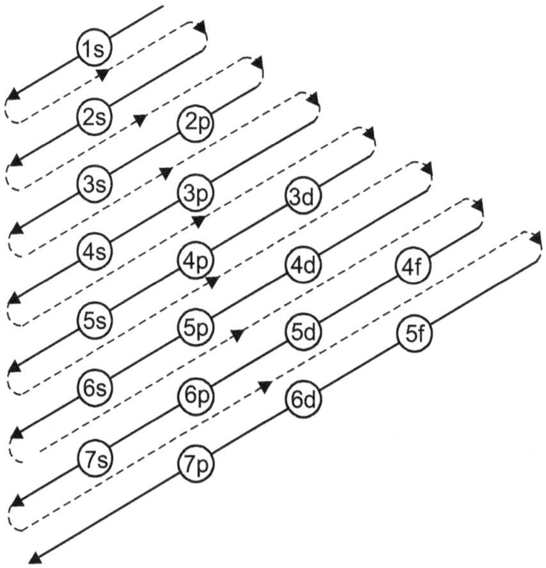

Fig. 2.6 : The approximate order in which the orbitals are filled with increasing number of electrons

2.5.3 Hund's Rule of Maximum Multiplicity

Once we understand clearly the filling order of orbitals, next question is how to fill a particular orbital with available number of electrons. For example, electron configuration of carbon is $1s^2$, $2s^2$, $2p^2$. The two electrons in 2p orbital may be accommodated by the following three ways :

$2p_x$	$2p_y$	$2p_z$
↑↓		

(i)

$2p_x$	$2p_y$	$2p_z$
↑	↓	

(ii)

$2p_x$	$2p_y$	$2p_z$
↑	↑	

(iii)

All the three arrangements follow Pauli exclusion principle, so we must determine which arrangement gives the greatest stability. To understand this we must follow Hund's rule, the essential parts of this rule may be stated as follows :

(a) Pairing of electrons will not start in any p, d and f orbitals until all the orbitals of that kind contain at least one electron each.

(b) In a set of degenerate orbitals, the electron preferentially occupies separate vacant orbitals first with parallel spin and when all the degenerate orbitals are singly occupied, then the pairing will start.

Thus, arrangement numbers (i) and (ii) are ruled out and **arrangement number (iii) is of low energy**. Similarly, for np^3 configuration, the correct arrangement is

↑	↑	↑

and not like

↑↓	↑	

(c) Hund's rule can be explained more correctly in terms of spectroscopic term, that is, multiplicities associated with electron configuration. The rules states that, **'an electron configuration with maximum spin multiplicity, represents the ground state of an atom'**. For example, electron configuration of nitrogen atom is $1s^2$, $2s^2$, $2p^3$, it can be represented as,

(i)

↑	↑	↑

or

(ii)

↑↓	↑	

Total spin momentum for these two arrangements is given by,

$$\vec{s} = ms_1 + ms_2 + ms_3 = +\frac{1}{2}+\frac{1}{2}+\frac{1}{2} = +\frac{3}{2}$$

$$\vec{s} = ms_1 + ms_2 + ms_3 = +\frac{1}{2}-\frac{1}{2}+\frac{1}{2} = +\frac{1}{2}$$

$$\text{Spin multiplicity} = (2s + 1) = \left(2 \times \frac{3}{2} + 1\right) = 4$$

$$\text{Spin multiplicity} = (2s + 1) = \left(2 \times \frac{1}{2} + 1\right) = 2$$

As the most stable configuration of an atom corresponds to maximum spin multiplicity, therefore configuration (i) is most correct one.

2.6 ELECTRONIC CONFIGURATION OF ELEMENTS

Based upon the principles discussed above, it is possible now to write the ground state electronic configuration of first few elements. While writing the electron configuration, we should keep in mind that, the number of electrons in a neutral atom is equal to its atomic number.

The electron in a ground state hydrogen atom must be in the 1s orbital, so its electronic configuration is written as $1s^1$, which denotes following things :

$1s^1$

- 1 (coefficient): Denotes the principal quantum number, n
- s: Denotes the subshell and the angular momentum quantum number, *l*
- superscript 1: Denotes the number of electrons in the orbital or subshell

The electron can also be represented by an orbital diagram that also shows the spin of the electron.

H = [↑]
$1s^1$

where the box [ ] denotes single (s) orbital and upward arrow indicates one of the two possible spins of the electron. Helium has two electrons, the second electron also occupies 1s level. This is possible because two electrons can have opposite spins (so different values of $\left(m_s, +\frac{1}{2} \text{ and } -\frac{1}{2}\right)$. This 1s level is now full.

He = [↑↓]
$1s^2$

The next atom lithium has three electrons. The third electron occupies the next lowest level, that is 2s level. So lithium will have electron configuration $1s^2, 2s^1$.

The fourth electron in Beryllium also occupies 2s level. Boron must have its fifth electron in the 2p level as the 2s level is now full. The sixth electron in carbon is also in the 2p level. As per Hund's rule the number of unpaired electrons in a given level is maximum, thus, in the ground state the two p electrons in carbon are unpaired. They occupy separate p orbitals and have parallel spin. Similarly, in nitrogen the three p electrons are unpaired and have parallel spins. From oxygen onwards pairing of electrons in p orbital will start. The electronic configuration for first ten elements can be represented as follows.

Element	Z		1s	2s	2p		
H	(1)	:	↑ ($1s^1$)				
He	(2)	:	↑↓ ($1s^2$)				
Li	(3)	:	↑↓ ($1s^2$)	↑ ($2s^1$)			
Be	(4)	:	↑↓ ($1s^2$)	↑↓ ($2s^2$)			
B	(5)	:	↑↓ ($1s^2$)	↑↓ ($2s^2$)	↑		($2p^1$)
C	(6)	:	↑↓ ($1s^2$)	↑↓ ($2s^2$)	↑	↑	($2p^2$)
N	(7)	:	↑↓ ($1s^2$)	↑↓ ($2s^2$)	↑	↑	↑ ($2p^3$)
O	(8)	:	↑↓ ($1s^2$)	↑↓ ($2s^2$)	↑↓	↑	↑ ($2p^4$)
F	(9)	:	↑↓ ($1s^2$)	↑↓ ($2s^2$)	↑↓	↑↓	↑ ($2p^5$)
Ne	(10)	:	↑↓ ($1s^2$)	↑↓ ($2s^2$)	↑↓	↑↓	↑↓ ($2p^6$)

Similarly, we may write the electronic configuration of elements having atomic numbers 11 to 20. Following Table 2.3 shows electronic configuration of elements with atomic numbers 11 to 20.

Table 2.3 : Electronic configuration of elements with Z = 11 – 20

Element	Symbol	Atomic number	Electronic configuration
Sodium	Na	11	$1s^2, 2s^2, 2p^6, 3s^1$
Magnesium	Mg	12	$1s^2, 2s^2, 2p^6, 3s^2$
Aluminium	Al	13	$1s^2, 2p^2, 2p^6, 3s^2, 3p_x^1$
Silicon	Si	14	$1s^2, 2s^2, 2p^6, 3s^2, 3p_x^1, 3p_y^1$
Phosphorous	P	15	$1s^2, 2s^2, 2p^6, 3s^2, 3p_x^1, 3p_y^1, 3p_z^1$
Sulphur	S	16	$1s^2, 2s^2, 2p^6, 3s^2, 3p_x^2, 3p_y^1, 3p_z^1$
Chlorine	Cl	17	$1s^2, 2s^2, 2p^6, 3s^2, 3p_x^2, 3p_y^2, 3p_z^1$
Argon	Ar	18	$1s^2, 2s^2, 2p^6, 3s^2, 3p_x^2, 3p_y^2, 3p_z^2$
Potassium	K	19	$1s^2, 2s^2, 2p^6, 3s^2, 3p^6, 4s^1$
Calcium	Ca	20	$1s^2, 2s^2, 2p^6, 3s^2, 3p^6, 4s^2$

It may be noted here that in case of potassium (K, 19) and calcium (Ca, 20), 3p orbital is followed by 4s and not 3d since energy of 4s is less than 3d [n + *l* rule, 4s (4 + 0 = 4) and 3d (3 + 2 = 5)].

Calcium onwards that is after filling of 4s level next available orbital is 3d and not 4p because both have n + *l* value = 5 (3d, 3 + 2 = 5, and 4p, 4 + 1 = 5) but for 3d, n value is less than for 4p and so 3d fills first. Filling of 3d orbital starts from element scandium (Z = 21) and is continued till element zinc (Z = 30). Following Table 2.4 shows electronic configuration of elements with atomic numbers 21 to 30.

Table 2.4 : Electronic configuration of elements with Z = 21 – 30

Element	Symbol	Atomic number 'Z'	Electronic configuration
Scandium	Sc	21	[Ar] $4s^2$ $3d^1$
Titanium	Ti	22	[Ar] $4s^2$ $3d^2$
Vanadium	V	23	[Ar] $4s^2$ $3d^3$
Chromium	Cr	24	**[Ar] $4s^1$ $3d^5$**
Manganese	Mn	25	[Ar] $4s^2$ $3d^5$
Iron	Fe	26	[Ar] $4s^2$ $3d^6$
Cobalt	Co	27	[Ar] $4s^2$ $3d^7$
Nickel	Ni	28	[Ar] $4s^2$ $3d^8$
Copper	Cu	29	**[Ar] $4s^1$ $3d^{10}$**
Zinc	Zn	30	[Ar] $4s^2$ $3d^{10}$

The configuration of elements are represented by a noble gas core, which shows in brackets the noble gas element that nearly precedes the element. For example, the electronic configuration of scandium (Z = 21) is given as $1s^2$, $2s^2$, $2p^6$, $3s^2$, $3p^6$, $4s^2$, $3d^1$. Since $1s^2$, $2s^2$, $2p^6$, $3s^2$, $3p^6$ is the electronic configuration of argon, we can simply write the electronic configuration of scandium by writing [Ar] $4s^2$, $3d^1$, where [Ar] denotes the 'Argon core'.

2.6.1 Anomalous Configuration of Chromium and Copper

The elements from scandium (Z = 21) to zinc (Z = 30) form first transition series. In this series, the electrons are added in the 3d orbitals according to Hund's rule. However, there are two irregularities. **The electronic configuration of chromium (Z = 24) is [Ar] $4s^1$ $3d^5$ and not [Ar] $4s^2$ $3d^4$ as we expect**. A similar irregularity is observed for copper, whose electronic configuration is **[Ar] $4s^1$ $3d^{10}$** rather than [Ar] $4s^2$ $3d^9$. This anomalous behavour is due to closeness of the 3d and 4s orbital energies. At the beginning of the series, the 4s level has a lower energy than the 3d level because it is more penetrating. However, as we move from left to right (Sc to Zn), the nuclear charge increases, the Argon core of electrons is pulled closer toward the nucleus and penetration effect diminishes. Consequently, the 3d level becomes more and more stable compared to 4s level. In addition to this, there is slightly greater stability associated with the half filled ($3d^5$) and completely filled ($3d^{10}$) subshells. Thus, instead of $3d^4$ or $3d^9$, the more stable configurations are $3d^5$, $3d^{10}$.

According to Hund's rule, the orbital diagram for Cr is,

Cr (24) → [Ar] [↑] $4s^1$ [↑ | ↑ | ↑ | ↑ | ↑] $3d^5$

Similarly, the orbital diagram for copper is,

Cu (29) → [Ar] [↑] $4s^1$ [↑↓ | ↑↓ | ↑↓ | ↑↓ | ↑↓] $3d^{10}$

The electronic configurations of elements with atomic number 31 onwards can be written on similar lines following general rules and also considering the extra stability associated with half filled and completely filled subshell configuration. Following are the examples in which the actual electronic configurations are different than that expected as per Aufbau principle due to the extra stability caused because of half filled or completely filled 'd' orbitals.

Mo (42) – [Kr] $5s^1$ $4d^5$

Pd (46) – [Kr] $4d^{10}$

Ag (47) – [Kr] $5s^1$ $4d^{10}$

Au (79) – [Kr] $6s^1$ $4f^{14}$ $5d^{10}$

Following Table 2.5 shows electronic configurations of elements from atomic number 31 onwards.

Table 2.5 : The electronic configuration of elements

Atomic number Z	Symbol	Electronic configuration	Atomic number Z	Symbol	Electronic configuration
31	Ga	[Ar] $4s^2$ $3d^{10}$ $4p^1$	71	Lu	[Xe] $6s^2$ $4f^{14}$ $5d^1$
32	Ge	[Ar] $4s^2$ $3d^{10}$ $4p^2$	72	Hf	[Xe] $6s^2$ $4f^{14}$ $5d^2$
33	As	[Ar] $4s^2$ $3d^{10}$ $4p^3$	73	Ta	[Xe] $6s^2$ $4f^{14}$ $5d^3$
34	Sc	[Ar] $4s^2$ $3d^{10}$ $4p^4$	74	W	[Xe] $6s^2$ $4f^{14}$ $5d^4$
35	Br	[Ar] $4s^2$ $3d^{10}$ $4p^5$	75	Re	[Xe] $6s^2$ $4f^{14}$ $5d^5$
36	Kr	[Ar] $4s^2$ $3d^{10}$ $4p^6$	76	Os	[Xe] $6s^2$ $4f^{14}$ $5d^6$
37	Rb	[Kr] $5s^1$	77	Ir	[Xe] $6s^2$ $4f^{14}$ $5d^7$
38	Sr	[Kr] $5s^2$	78	Pt	[Xe] $6s^1$ $4f^{14}$ $5d^9$

Atomic number Z	Symbol	Electronic configuration	Atomic number Z	Symbol	Electronic configuration
39	Y	[Kr] $5s^2$ $4d^1$	79	Au	[Xe] $6s^1$ $4f^{14}$ $5d^{10}$
40	Zr	[Kr] $5s^2$ $4d^2$	80	Hg	[Xe] $6s^2$ $4f^{14}$ $5d^{10}$
41	Nb	[Kr] $5s^1$ $4d^4$	81	Tl	[Xe] $6s^2$ $4f^{14}$ $5d^{10}$ $6p^1$
42	Mo	[Kr] $5s^1$ $4d^5$	82	Pb	[Xe] $6s^2$ $4f^{14}$ $5d^{10}$ $6p^2$
43	Tc	[Kr] $5s^2$ $4d^5$	83	Bi	[Xe] $6s^2$ $4f^{14}$ $5d^{10}$ $6p^3$
44	Ru	[Kr] $5s^1$ $4d^7$	84	Po	[Xe] $6s^2$ $4f^{14}$ $5d^{10}$ $6p^4$
45	Rh	[Kr] $5s^1$ $4d^8$	85	At	[Xe] $6s^2$ $4f^{14}$ $5d^{10}$ $6p^5$
46	Pd	[Kr] $4d^{10}$	86	Rn	[Xe] $6s^2$ $4f^{14}$ $5d^{10}$ $6p^6$
47	Ag	[Kr] $5s^1$ $4d^{10}$	87	Fr	[Rn] $7s^1$
48	Cd	[Kr] $5s^2$ $4d^{10}$	88	Ra	[Rn] $7s^2$
49	In	[Kr] $5s^2$ $4d^{10}$ $5p^1$	89	Ac	[Rn] $7s^2$ $6d^1$
50	Sn	[Kr] $5s^2$ $4d^{10}$ $5p^2$	90	Th	[Rn] $7s^2$ $6d^2$
51	Sb	[Kr] $5s^2$ $4d^{10}$ $5p^3$	91	Pa	[Rn] $7s^2$ $5f^2$ $6d^1$
52	Te	[Kr] $5s^2$ $4d^{10}$ $5p^4$	92	U	[Rn] $7s^2$ $5f^3$ $6d^1$
53	I	[Kr] $5s^2$ $4d^{10}$ $5p^5$	93	Np	[Rn] $7s^2$ $5f^4$ $6d^1$
54	Xe	[Kr] $5s^2$ $4d^{10}$ $5p^6$	94	Pu	[Rn] $7s^2$ $5f^6$
55	Cs	[Xe] $6s^1$	95	Am	[Rn] $7s^2$ $5f^7$
56	Ba	[Xe] $6s^2$	96	Cm	[Rn] $7s^2$ $5f^7$ $6d^1$
57	La	[Xe] $6s^2 5d^1$	97	Bk	[Rn] $7s^2$ $5f^9$
58	Ce	[Xe] $6s^2$ $4f^1$ $5d^1$	98	Cf	[Rn] $7s^2$ $5f^{10}$
59	Pr	[Xe] $6s^2$ $4f^3$	99	Es	[Rn] $7s^2$ $5f^{11}$
60	Nd	[Xe] $6s^2$ $4f^4$	100	Fm	[Rn] $7s^2$ $5f^{12}$
61	Pm	[Xe] $6s^2$ $4f^5$	101	Md	[Rn] $7s^2$ $5f^{13}$
62	Sm	[Xe] $6s^2$ $4f^6$	102	No	[Rn] $7s^2$ $5f^{14}$
63	Eu	[Xe] $6s^2$ $4f^7$	103	Lr	[Rn] $7s^2$ $5f^{14}$ $6d^1$
64	Gd	[Xe] $6s^2$ $4f^7$ $5d^1$	104	Unq	[Rn] $7s^2$ $5f^{14}$ $6d^2$
65	Tb	[Xe] $6s^2$ $4f^9$	105	Unp	[Rn] $7s^2$ $5f^{14}$ $6d^3$
66	Dy	[Xe] $6s^2$ $4f^{10}$	106	Unh	[Rn] $7s^2$ $5f^{14}$ $6d^4$
67	Ho	[Xe] $6s^2$ $4f^{11}$	107	Uns	[Rn] $7s^2$ $5f^{14}$ $6d^5$
68	Er	[Xe] $6s^2$ $4f^{12}$	108	Uno	[Rn] $7s^2$ $5f^{14}$ $6d^6$
69	Tm	[Xe] $6s^2$ $4f^{13}$	109	Une	[Rn] $7s^2$ $5f^{14}$ $6d^7$
70	Yb	[Xe] $6s^2$ $4f^{14}$	110	Uun	[Rn] $7s^2$ $5f^{14}$ $6d^8$

The symbol [Ar] is called Argon core and it represents $1s^2$, $2s^2$, $2p^6$, $3s^2$, $3p^6$. [Kr] is called Krypton core and it represents [Ar] $4s^2$ $3d^{10}$ $4p^6$, [Xe] is called the Xenon core and it represents [Kr] $5s^2$ $4d^{10}$ $5p^6$ while [Rn] is called radon core and it means [Xe] $6s^2$ $4f^{14}$ $5d^{10}$ $6p^6$.

2.7 LONG FORM OF THE PERIODIC TABLE

The long form of the periodic table is based upon the electronic configuration of elements. It is present form or the modern periodic table of elements.

2.7.1 Modern Periodic Table

The properties of an element are largely governed by the number of electrons in the outer shell and their arrangement. If we arrange the elements in groups which have same outer electronic arrangement, then elements within the group should show similarities in properties. One great advantage of this is that initially it is only necessary to learn the properties of each group rather than the properties of each individual element. Each group thus contains elements of similar electronic configuration. For example, all group 1 elements have an outer electron that is ns^1, where n is the principal quantum number. These elements thus show many similarities in physical and chemical properties.

The name periodic table is derived from the fact that the properties of the elements which are tabulated in order of increasing atomic number are periodic. That is **properties of elements reoccur after certain interval of atomic number. This is known as periodic law**. The ideal periodic arrangement should be based on the electronic configuration. Thus, periodic table is simply a device for listing the elements so that elements with similar properties are grouped together.

Modern periodic table is the long form of periodic table. A horizontal row in a periodic table is called **period** and vertical row is called **group**. **There are 18 vertical columns which define chemical families or groups**. There are 7 periods in the long form of periodic table and these are numbered from 1 to 7. In the first period, there are only two elements whereas in the second period there are eight elements. Table 2.6 shows the number of elements present in each period.

Table 2.6 : Number of elements present in each period

Period	Orbitals to be filled	Number of elements in the period
1	1s	2
2	2s, 2p	8
3	3s, 3p	18
4	4s, 3d, 4p	18
5	5s, 4d, 5p	18
6	6s, 4f, 5d, 6p	32
7	7s, 5f, 6d, 7p	Remaining elements

Each period starts with the entry of electron in new shell. Thus, the general electronic configuration of the first element of every period is ns^1, where n = 1, 2, 3.... Similarly, the general electron configuration of the last element of every period is $ns^2\ np^6$ (except first period, where it is $2s^2$ of Helium).

Periodic table is expanded to accommodate more than 100 elements, listed by atomic numbers. There are 8 groups which are designated by letter A or B.

In the United States, the practice is to use A to designate the representative elements and B to designate the transition elements. (Refer Table 2.7).

In Europe, the tradition is to use B for the representative elements (after alkali and alkaline metals) and A for the transition elements.

To overcome confusion between A and B **IUPAC**, the **'International Union of Pure and Applied Chemistry'** has recommended use of Arabic numbers 1 to 18 as shown in Table 2.7.

Table 2.7 : Today's Periodic Table

LONG FORM OF PERIODIC TABLE

s-block or Light Metals | Heavy Metals or d-block (Transition Metals) | p-block or Non-Metals

IA - Alkali metal
IIA - Alkaline earth metal
- - - - Metalloids

	1 IA	2 IIA	3 IIIB	4 IVB	5 VB	6 VIB	7 VIIB	8 VIII	9 VIII	10 VIII	11 IB	12 IIB	14 IIIA	15 IVA	16 VA	17 VIA	18 VIIA	VIIA or 0
Period 1	1 H																	2 He
Period 2	3 Li	4 Be											5 B	6 C	7 N	8 O	9 F	10 Ne
Period 3	11 Na	12 Mg											13 Al	14 Si	15 P	16 S	17 Cl	18 Ar
Period 4	19 K	20 Ca	21 Sc	22 Ti	23 V	24 Cr	25 Mn	26 Fe	27 Co	28 Ni	29 Cu	30 Zn	31 Ga	32 Ge	33 As	34 Se	35 Br	36 Kr
Period 5	37 Rb	38 Sr	39 Y	40 Zr	41 Nb	42 Mo	43 Tc	44 Ru	45 Rh	46 Pd	47 Ag	48 Cd	49 In	50 Sn	51 Sb	52 Te	53 I	54 Xe
Period 6	55 Cs	56 Ba	57 to 71	72 Hf	73 Ta	74 W	75 Re	76 Os	77 Ir	78 Pt	79 Au	80 Hg	81 Ti	82 Pb	83 Bi	84 Po	85 At	86 Rn
Period 7	87 Fr	88 Ra	89 to 103	104 Rf	105 Ha	106 Sg	107 Ns	108 Hs	109 Mt									

f-block

Lanthanide series →	57 La	58 Ce	59 Pr	60 Nd	61 Pm	62 Sm	63 Eu	64 Gd	65 Tb	66 Dy	67 Ho	68 Er	69 Tm	70 Yb	71 Lu
Actinide series →	89 Ac	90 Th	91 Pa	92 U	93 Np	94 Pu	95 Am	96 Cm	97 Bk	98 Cf	99 Es	100 Fm	101 Md	102 No	103 Lr

2.8 PERIODIC CLASSIFICATION OF ELEMENTS

Over hundreds of elements are listed in periodic table (Table 2.7). According to type of subshell being filled, the elements can be classified into four types :

(i) Representive or main group elements.
(ii) Noble or inert gas elements.
(iii) Transition elements, and
(iv) Inner transition elements.

2.8.1 Representative or Main Group Elements

These elements are also called normal elements. These are elements in groups **1, 2, 13 to 17, all** of which have incompletely filled s or p subshells of highest principal quantum number 'n'. The general electronic configuration of these elements is ns^1, ns^2 and $ns^2\ np^1$ to ns^2, np^5. All the elements of a particular group of representive elements will have similar physical and chemical properties.

The elements of this type may loose the electrons, gain the electrons or share the electrons such that they attain the nearest inert gas electron configuration. For example, elements of alkali metal family loose the outermost electron while elements of halogen family tend to gain electron so that they acquire stability. The elements in between these two extremes, like carbon family, nitrogen family etc. generally share the electrons with other elements and acquire stability.

2.8.2 Noble or Inert Gas Elements

The members of this family are He, Ne, Ar, Kr, Xe and Rn. This family (group 18) is placed at the extreme right of the periodic table. With the exception of Helium, all other members have a completely filled p-subshell. General electronic configuration of these elements is $\mathbf{ns^2\ np^6}$ **(Helium – $\mathbf{1s^2}$)** where n = principal quantum number. As these elements have their outermost shell completely filled, they do not react readily with other elements and so called inert gases. They are also called as rare gases or aerogens.

2.8.3 Transition Elements

The transition metals are the elements in group 3 to 12. They have incompletely filled d subshells in one of the common oxidation states. The transition elements are also called as d-block elements. The general electronic configuration of these elements is $ns^{1\text{ or }2}$, $(n - 1)d^{1-10}$. All transition elements are metals. There are 3 complete and 4th incomplete series of transition elements corresponding to the filling of 3d, 4d, 5d and 6d orbitals. First transition series starts from Scandium (21) to Zinc (30), second starts from Yttrium (39) to Cadmium (48), third series includes Lanthanum (57), Hafnium (72) to Mercury (80) whereas 4th series starts from Actinium (89) and is incomplete.

Transition elements show remarkable similarities in their physical properties. These elements form large number of complexes which are coloured. Transition elements show variable oxidation states and find use in many catalytic processes.

2.8.4 Inner Transition Elements

These elements are placed at the bottom of periodic table. Actually they must be placed in group 3. In the atoms of these elements, the addition of successive electron takes place in **(n – 2) f shell**. General electronic configuration of these elements is $\mathbf{ns^2, (n-1)d^{0-1}, (n-2)f^{1-14}}$. There are two series of these elements, namely lanthanides (n = 6) and actinides (n = 7). In each series, there are fourteen elements.

Lanthanides (first series) are the elements that follow the element lanthanum (Z = 57). The series starts from Cerium (58) and ends with Lutetium (71). Actinides (second series) are the elements that follow the element Actinium (Z = 89). The series starts from Thorium (90) and ends with Lawrencium (103). These elements (lanthanides and actinides) are also called as f-block elements.

2.9 BLOCKS IN THE PERIODIC TABLE

On the basis of entry of differentiating electron (which is a last electron of an atom), periodic table is subdivided into four blocks. The blocks are **s, p, d and f**. Accordingly, elements are also classified as

(i) s-block elements, (ii) p-block elements,

(iii) d-block elements, and (iv) f-block elements.

For example, electronic configuration of sodium is given as $_{11}Na = 1s^2, 2s^2, 2p^6, 3s^1$. The differentiating electron of sodium enters in 3s orbital, hence, sodium is placed in 's' block of periodic table.

Following Fig. 2.7 shows various blocks in the periodic table.

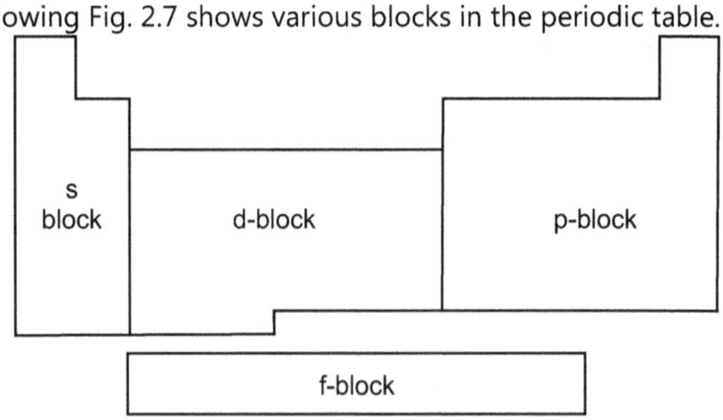

Fig. 2.7 : Various blocks in the periodic table

2.9.1 s-Block Elements

In these elements, differentiating electron enters the ns orbital which is being filled. The elements of groups 1 (alkali metals) and group 2 (alkaline earth metals) belong to this block. General electronic configuration of group-1 elements is ns^1, while that of group-2 elements is ns^2. This block is placed at the extreme left of the periodic table.

2.9.2 p-Block Elements

Thee elements in which p-orbitals are being progressively filled are called p-block elements. The elements of **groups −13 to 17 and 18** (exception of Helium whose electronic configuration is $1s^2$) are the members of p-block. Since, in the atoms of these elements the differentiating electron enters the p-orbital namely 2p, 3p, 4p, 5p and 6p etc., the characteristic configuration of these elements vary from **ns^2 np^1 (group 3) to ns^2, np^6 (group 18)**. This block is placed at the extreme right of the periodic table. This block consists of some metals, non-metals and metalloids. It also includes noble gases.

2.9.3 d-Block Elements

The elements in which the differentiating electron enters the **(n – 1) d orbital** are called d-block elements. These elements are placed at the middle part of the periodic table between s and p-block elements. The valence shell electronic configuration of the atoms of the d-block elements is given as **ns^{1-2}, $(n - 1)d^{1-10}$**, where (n – 1) represents penultimate (last but one) shell.

These elements which are also called as transition elements comprise of four series corresponding to the filling of 3d, 4d, 5d and 6d orbitals of $(n - 1)^{th}$ main shell. First three series contains ten elements each. The fourth series is incomplete.

2.9.4 f-Block Elements

The elements in which the (n – 2) f orbital of the atom is filled progressively are called f-block elements. The atoms of these elements have their general valence shell electronic configuration represented as ns^2, $(n - 1)d^{0-1}$, $(n - 2)f^{1-14}$. f-block elements are divided into two series, namely Lanthanides and Actinides. Both these series are of 14 elements each. This block is placed at the bottom of the periodic table.

2.10 PERIODIC LAW AND PERIODICITY IN PROPERTIES

After the discovery of large number of elements, it was necessary that there must be some order or regularity which control the properties of elements. Mendeleev in 1869 stated a law called periodic law, which was first successful attempt in the classification of elements and their comparative study. The law stated as follows : **'The physical and chemical properties of elements are the periodic function of their atomic weights'. Mendeleev** thus arranged the elements in the order of their increasing atomic weights in the form of table known as **Mendeleev's periodic table**. The table made by Mendeleev comprised of groups and periods. There were number of vacant places or gaps for the elements not known at that time. Mendeleev predicted their properties and when these elements were later on discovered, their properties remarkably matched with the predicted ones.

Afterwards scientists found many defects and shortcomings in the Mendeleev's periodic table. The classification based on atomic number could not stand for long time, because atomic weight is the nuclear property, which depends upon the number of nucleons that is number of protons and neutrons. The neutron number in an atom may vary and therefore atomic weight becomes variable property. Thus, it is of no use for the classification of elements. The physical and chemical properties of an atom of element clearly depend upon the number of electrons present, that is, on the atomic number. The atomic number is non-variable property. Thus, most of defects and anomalies in the Mendeleev's periodic table may be overcome if basis of classification is changed from atomic weight to atomic number. Hence, **Mendeleev's periodic law was modified, which is known as Modern periodic law. According to this law, 'the physical and chemical properties of elements are periodic function of their atomic numbers'.**

2.11 PERIODICITY IN PROPERTIES

In a particular group of periodic table, certain physical and chemical properties are reoccurring. This **reoccurrence of properties of elements after a regular interval is called 'periodicity'**. The interval here is the period. The physical and chemical properties of element depend on arrangement of electrons in the valency shell, that is, electronic configuration of atom. So the elements with similar arrangement of electrons in valence shell (outermost shell) will show similar properties. The property observed for one element of a family reappears for the second element of the same family after passing through a certain interval, that is, period. For example, sodium ($3s^1$, member of group 1, third period) and potassium ($4s^1$, member of group 1, fourth period) will show similar properties because both of them have similar number and arrangement of electrons in the outermost shell. There are many properties such as atomic radius, ionic radius, atomic volume, density, ionization, energies, electronegativity etc. which are periodic.

There can be more or less regular change in the similar properties as we go from top to bottom in a particular group. The similarities in property are due to similar valence shell electronic configuration of the members of family. The gradual change in properties is due to difference in number of electrons in penultimate and antipenultimate shells, as well as gradual change in volume of the element in a particular family.

In the beginning of this topic we studied that the electronic configuration of the elements vary periodically and therefore physical and chemical properties of these elements also vary periodically with increasing atomic number. Such properties are called periodic properties. Some physical properties of elements are discussed in the following sections.

2.11.1 Size of Atoms and Ions (Atomic and Ionic Radii)

Size of atoms and ions are generally described in terms of atomic radius. Physical properties such as density, melting point and boiling point are related to atomic size, but atomic size is difficult to define.

In metallic elements, the atomic radius is one half the distance between the two nuclei in two adjacent metal atoms. Fig. 2.8 shows the distance between two adjacent metal atoms (d).

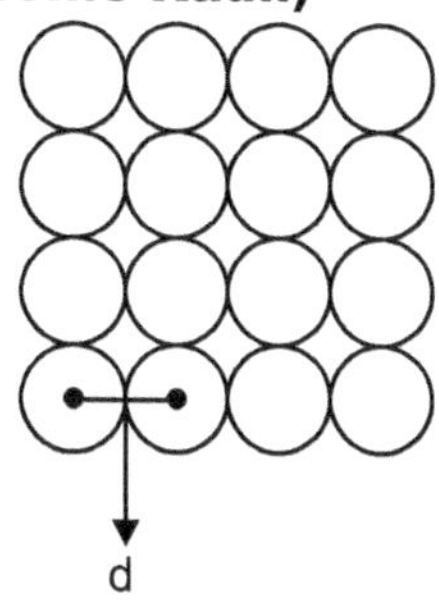

Fig. 2.8

Atomic radius is one half the distance between the centre of two adjacent atoms $\mathbf{r} = \frac{1}{2}\mathbf{d}$, where r = atomic radius (size), d = distance between the centres of two adjacent atoms, that is, internuclear distance.

For the elements that exist as simple diatomic molecules, the atomic radius is one-half the distance between the nuclei of the two atoms in a particular molecule. Fig. 2.9 shows simple diatomic molecule.

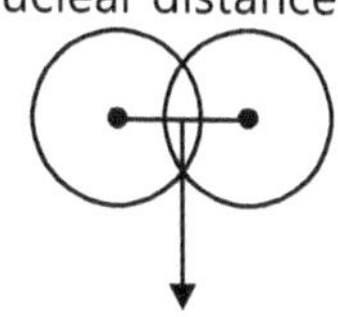

Fig. 2.9

(A) Atomic size :

Atomic radius is determined to a large extent by how strongly the outer shell electrons are held by the nucleus. The atomic radius depend upon three factors : (i) nuclear charge, (ii) number of shells and (iii) screening effect. As the nuclear charge increases, electrons are pulled towards nucleus (inward) and size is found to decrease. With increase in number of shells, electrons go more and more away from the nucleus and size is observed to increase. More is the number of inner electrons, screening effect will be more, which will result in increase in atomic size. These three trends are important while studying the periodic trends of atomic size across the period and in the group.

Across the period as we move from left to right (in case of s and p block), the atomic radius goes on decreasing. This is because as we move from left to right, the effective nuclear charge increases and electrons are added in the same shell. With increase in number of electrons screening effect also increases but increase in charge is more effective than the increase in screening effect and therefore atomic radius decreases steadily across the period. For example, consider the third period, that is, from sodium to argon.

Following Table 2.8 shows atomic radius of 3rd row (period) elements.

Table 2.8 : Atomic radius of 3rd row (period) elements

Atoms	Na	Mg	Al	Si	P	S	Cl	Ar
Atomic radius (pm)	186	160	143	132	128	127	99	98

Picometer = pm = 10^{-12} m = 100 pm = 1 A°.

In the group (down the family), it is observed that atomic radius increases with increasing atomic number. For example, in group 1 (alkali metal family), the valence electrons are in ns orbital and since orbital size increases with the increase in principal quantum number n, the size of atoms is found to increase from Li to Cs, as we go down the group.

Table 2.9 shows atomic radius of group 1 metals.

Table 2.9 : Atomic radius of 1st group elements

Atoms	**Atomic radius (pm)**
Li (3)	152
Na (11)	186
K (19)	227
Rb (37)	248
Cs (55)	265

As we go down the group (top to bottom), the atomic number increases, hence nuclear charge increases, extra electrons are added in a new shell and screening effect increases. The increase in screening effect and addition of a new shell becomes more effective than increase in nuclear charge. Therefore, atomic radius is found to increase down the group by s and p block elements.

(B) Ionic size :

Ionic size is the radius of a cation or anion. Ionic radius affects the physical and chemical properties of an ionic compound. When a neutral atom is converted to an ion, there is change in size. If the atom forms an ion (−ve ion) by taking one or more electrons, its size is found to increase. This is because net number of negative charges (that is, number of electrons) are more than net number of positive charges (that is, number of protons), hence repulsive forces between the electrons are stronger than attractive forces between the electrons and protons. The net result is increase in size of anion than parent atom.

On the other hand, a cation (+ve ion) is smaller than neutral atom, because cation is formed by removing one or more electrons from the neutral atom. While doing so, sometimes one shell is completely removed, not only that, in a cation net number of negative charges are less than net number of positive charges, hence repulsive forces between the electrons are weaker than attractive forces between the electrons and protons. Because of stronger attractive forces, electrons are pulled towards nucleus thereby decreasing size.

Following Table 2.10 shows comparison between sizes of atoms and their ions.

Table 2.10 : Comparative sizes of atoms and ions

Size of atoms		Size of ions		Size of atoms		Size of ions	
Symbol	**pm**	**Symbol**	**pm**	**Symbol**	**pm**	**Symbol**	**pm**
Li	152	Li^{+}	60	F	72	F^{-}	136
Na	186	Na^{+}	95	Cl	99	Cl^{-}	181
K	227	K^{+}	133	Br	114	Br^{-}	195
Mg	160	Mg^{++}	65	I	133	I^{-}	216
Ca	197	Ca^{++}	99	O	73	O^{2-}	140
Al	143	Al^{+++}	50	S	127	S^{2-}	184
Ga	135	Ga^{+++}	62	N	92	N^{3-}	171

The general trend regarding ionic size is very similar to atomic size. In a group (top to bottom) ionic radius is found to increase while in a given period (left to right) ionic radius is found to decrease. The reason being the same as given for sizes of atoms.

Covalent radius : The covalent radius, r_{cov}, is a measure of the size of an atom that forms part of covalent bond. It is usually measured either in picometers or angstroms, with 1 A° = 100 pm. In principle, the sum of the two co equal the covalent bong length between two atoms, R = r + r.

Crystal radius : Ionic radius, r_{ion} is the radius of an atom's ion in ionic crystal structure. Although neither atoms nor ions have sharp boundaries. They are sometimes treated as if they were hard spheres with radii such that the sum of ionic radii of cation and anion gives the distance between ions in a crystal lattice.

2.11.2 Ionization Energy

It is the minimum energy required to remove an electron from a gaseous atom in its ground state. The magnitude of ionization energy is a measure of effort required to force an atom to give up an electron or it tells us how 'tightly' the electron is held in the atom. The higher the ionization energy, the more difficult is to remove the electron from the atom. Ionization energy is expressed in kJ $mole^{-1}$.

In an atom, the energy required to remove the first electron from the gaseous atom to form a unipositive gaseous ion is termed as first ionization energy (I_1).

$$X_{(g)} + \text{Energy} \longrightarrow X^{+}_{(g)} + e^{-} \qquad \text{First ionization } (I_1)$$

The second ionization potential (I_2) is the energy required to remove the electron from the monopositive ion, similarly there may be 3rd, 4th (I_3, I_4 ......) ionization potentials and so on.

$$X^{+}_{(g)} + \text{Energy} \longrightarrow X^{2+}_{(g)} + e^{-} \qquad \text{Second ionization } (I_2)$$

$$X^{++}_{(g)} + \text{Energy} \longrightarrow X^{3+}_{(g)} + e^{-} \qquad \text{Third ionization } (I_3)$$

After removing one electron from an atom, it gives ion (cation) which is small in size than parent atom. Number of electrons are less than number of protons and so electrons are firmly held by nucleus. Thus, if now we want to remove one more electron from an ion (+ve ion, cation), we require more energy. Therefore, $I_1 < I_2 < I_3 < ......$

Energy absorbed by an atom or an ion in the ionization process has a positive value. Thus, ionization energies are all positive quantities.

Following Table 2.11 shows variation of ionization energy across the period.

Table 2.11 : The ionization energies of second period element

Atoms		Li	Be	B	C	N	O	F	Ne
Ionization energy, kJ/mol									
First IE	I_1	520	899	801	1086	1400	1314	1680	2080
Second IE	I_2	7300	1757	2430	2350	2860	3390	3370	3950
Third IE	I_3	11815	14850	3660	4620	4580	5300	6050	6120

Apart from the small irregularities, **the ionization energies of element in a period increase with increasing atomic number**. As we move from left to right in a given period, the effective nuclear charge increases with increase in atomic number. The electrons are added in the same shell, thus, electrons are more close to the nucleus, hence energy required to remove electron is more. Larger the nuclear charge and small atomic size means electrons are more tightly held by the nucleus. Noble gases have highest of the ionization energy (Table 2.11, Neon) among the elements in the same period, this is because noble gas elements have special stability due to completed octet, and hence energy required to remove electron from the stable arrangement is highest. Similarly, ionization energies of group-1 elements (alkali metal family) seems to be lowest (Table 2.11, Lithium) this is because these elements have only one valence electron which is properly protected from the nuclear charge by inner electrons and thus it is less firmly held by the nucleus. In addition to this, after loosing that valence electron, alkali metals acquire inert gas configuration and become stable, thus they readily loose electron and hence energy required to remove electron is less.

The group-2 elements, that is, alkaline earth metals will have comparatively higher ionization energies than the alkali metals (group 1). The alkaline earth metals have two valence electrons (ns^2). These two s electrons do not shield each other properly and the effective nuclear charge for alkaline earth metal atom is larger than that for alkali metal atoms. Thus, energy required to remove the electron is higher.

Although in general the ionization energy increases from left to right in a given period, some irregularities are observed. First irregularity is between Beryllium and Boron. Be is a member of group-2 (ns^2) while Boron is a member of group-13 (ns^2, np^1). Ionization energy of B is unexpectedly less than Be, this is because group-13 elements (B) has single electron in the outermost p orbital, which is well shielded by the inner electrons. Therefore, less energy is needed to remove a single p electron than to remove the paired s electron from the sample principal energy level.

The second irregularity is observed between groups 15 and 16, that is between nitrogen and oxygen. Here also, ionization energy of oxygen is unexpectedly less than nitrogen. In group 15, that is, in the case of nitrogen ($2s^2$, $2p^3$) the p electrons are in three separate orbitals, while in group 16, that is, in the case of oxygen ($2s^2$, $2p^4$) the additional p electron must be paired with one of the three p electrons. This is shown below as,

N(7)	↑↓	↑	↑	↑	O(8)	↑↓	↑↓	↑	↑
	$2s^2$	$2p^3$				$2s^2$	$2p^4$		

The situation, where two electrons in the same orbital results in greater electrostatic repulsion, which make it easier to ionise an atom of group 16 element even though nuclear charge has increased by one unit. Thus, ionization energy of oxygen (group 16) is less than nitrogen (group 15).

In a particular group, ionization energy is found to decrease as we move down the group (top to bottom). This can be explained on the basis that, the elements in the same group have similar valence shell configuration, however as the principal quantum number 'n' increases down the group, the average distance of outer electrons and nucleus increases. The greater separation between the outer electron and nucleus means a weaker attractive forces, so that energy required to remove the electron is less and less as we go down the group. This clearly indicates that ionization energy is directly related to atomic size, smaller is the atom higher is the ionization energy and vice-versa. For example, ionization energies of group-1 metals are listed below in Table 2.12.

Table 2.12 : Ionization energies of group-1 metals

Atoms	Ionization energy, kJ/mol
Li	520.1
Na	495.7
K	418.6
Rb	402.9
Cs	375.6

In general, it is observed that **metals have relatively low ionization energies, whereas non-metals possess much higher ionization energies**. The ionization energies of the metalloid are in between those of metals and non-metals. The difference in ionization energies clearly suggests why metals always form cation (that is give up electrons) and non-metals form anions (that is gain of electrons) in ionic compounds.

2.11.3 Pauling's Electronegativity Scale

Whenever two elements are bonded by a covalent bond, both of them share a pair of electrons. The tendency of the bonded atom in a molecule to attract this shared pair of electron is termed as electronegativity. It differs from electron affinity since it is related with an atom in a molecule and not with isolated atom. Same atom in a different chemical environment may not have similar electronegativity. Therefore, the power of carbon atom to attract electron pair in CH_4 molecule will be more than that in CCl_4 molecule. Thus, electronegativity is not the fundamental property, but a relative property.

The widely accepted scale to measure the electronegativity values is Pauling's scale.

Following Table 2.13 shows values of electronegativity of some elements (on Pauling's scale).

Table 2.13 : The electronegativities of common elements

Group No.	1	2	13	14	15	16	17
	H 2.1						
	L 1.0	Be 1.5	B 2.0	C 2.5	N 3.0	O 3.5	F 4.0
	Na 0.9	Mg 1.2	Al 1.5	Si 1.8	P 2.1	S 2.5	Cl 3.0
	K 0.8	Ca 1.0	Ga 1.6	Ge 1.8	As 2.0	Se 2.4	Br 2.8
	Rb 0.8	Sr. 1.0	In 1.7	Sn 1.8	Sb 1.9	Te 2.1	I 2.5
	Cs 0.7	Ba 0.9	Tl 1.8	Pb 1.9	Bi 1.9	Po 2.0	At 2.2

It is seen from Table 2.13 that, the electronegativity is a periodic function of atomic number of elements. **In a period, it increases on moving from left to right whereas in a group it steadily decreases down the group**. Thus, electronegativity is related to atomic size. As atomic size increases electronegativity decreases and vice-versa.

Fluorine is the most electronegative element having maximum value equal to 4.0, which is at the top right hand corner of the periodic table. Caesium is the least electronegative element having lowest value equal to 0.7, which is at the bottom of left hand corner of the periodic table.

2.11.4 Shielding Effect and Shielding Constant

An atom of the element comprises of positively charged nucleus which attracts the electrons (which are revolving around the nucleus) towards itself. This kind of attractive forces between the nucleus and electrons are like that between a magnet and pins. Here nucleus acts as magnet and electrons like pins. The distance between nucleus and electrons decides the attractive forces between them. If electron is away from the nucleus it suffers less pull towards the nucleus. In a multielectron atom, we can divide the electrons into two types : one, electrons in the valence shells (that is, outermost shell, n) and two, electrons in penultimate (n – 1) and antipenultimate (n – 2) shells. Fig. 2.10 shows an electron picture of multielectron atom.

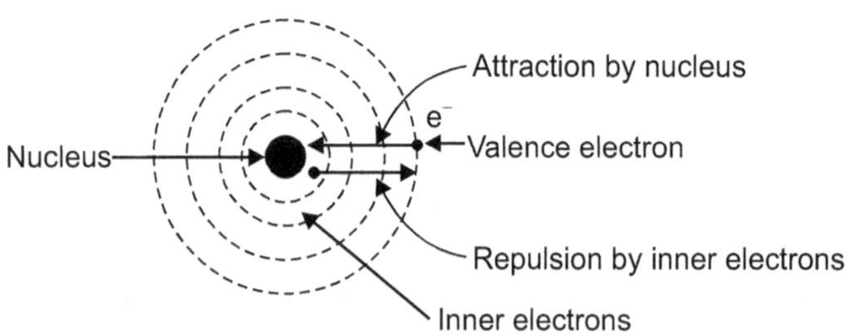

Fig. 2.10 : Structure of an atom, valence electrons and inner electrons

The electrons in outer shell (valence shell) suffer less pull by nucleus for the two reasons :

1. They are comparatively away from the nucleus and
2. In between nucleus and valence electron there are inner electrons which protect them from nuclear charge. This **decrease in attractive forces between the nucleus and the electrons due to inner**

electrons is known as screening or shielding effect. Here, the word shield is used for the role of inner electrons that they play for protecting outer electrons from the nuclear charge by coming between them.

Thus, large is the number of inner electrons (more number of inner shells), less is the attractive force acting on the valence electron by nucleus. This means strong shielding of outer electrons by inner electrons from the nuclear charge. The valence electrons therefore do not experience that pull because of total positive charge on the nucleus. The actual pull is due to reduced charge by number of inner electrons. **This reduced charge is called effective nuclear charge**. This effective nuclear charge can be calculated using following relation :

$$Z_{eff} \text{ or } Z^* = Z - S$$

where, Z_{eff} or Z^* = The effective nuclear charge or effective atomic number

Z = Actual nuclear charge or actual atomic number

S = The shielding or screening constant

Thus, to calculate effective nuclear charge, we need to know screening constant 'S'. The value of 'S' depends upon number and type of the inner electrons. Here, type refers to electrons in various subshells, such as, s-electrons, p-electrons and so on. Contribution by these electrons towards screening is not same. s-electrons are very good screening electrons, because they revolve around the nucleus in a spherical path and so they remain exactly between the outer electrons and nucleus for a maximum time. p-electrons are comparatively poor screening electrons because of their revolving path. They tend to go away from the nucleus while revolving around it. The time they spend between the nucleus and outer electron is therefore less. The contribution of p-electrons towards screening is thus less as compared to s-electrons.

On the other hand, d and f electrons are more diffuse and directional. Most of the time they are away from the nucleus while revolving around it. The time they spend between outer electrons and nucleus is thus lowest and their contribution towards screening is much less as compared to s and p-electrons. The order of shielding is thus,

$$s > p > d > f$$

For the calculation of screening constant, S, we need to count the number of inner electrons and their contribution towards screening. Using basic concepts and the Slater empirical rules, it is possible to calculate Z*, the effective nuclear charge.

Slater's Empirical Rules :

1. Write the electronic configuration of the element in shellwise manner, grouping s and p orbitals together as (1s) (2s, 2p) (3s 3p) (3d) (4s, 4p) (4d) (4f) (5s, 5p) etc.
2. Electrons in any group lying above that of a particular group contribute nothing to the screening constant. For example, in Nitrogen ($1s^2$, $2s^2$, $2p^3$), 2s and 2p electrons do not contribute to the screening constant of 2s group.
3. All the electrons in the same group shield a particular electron to an extent of 0.35 each. If it happens to 1s electron in 1s group, it contributes only to the extent of 0.30.
4. For s and p electrons, all the electrons in the (n – 1) shell shield to the extent of 0.85 each.
5. For s and p electrons in (n – 2) or lower shells shield completely that is their contribution is 1.00 each.
6. When the electron being shielded is a d or f electron then all the electrons lying below it in a group contribute to the extent of 1.00 each.

Example : Let us calculate the value of Z* for the valence electron of Aluminium.

Solution : Aluminium has Z = 13 and its electronic configuration is $1s^2$, $2s^2$, $2p^6$, $3s^2$, $3p^1$.

We are interested in finding the screening constant of 13th electron, that is, $3p^1$ electron, of Aluminium. Hence, using Slater's rule, we proceed as, we write the grouping of the orbitals following rule 1 as, $(1s^2)$ $(2s^2, 2p^6)$ $(3s^2, 3p^1)$.

Using rules 3, 4 and 5, we may find the contribution of each electron toward screening as,

$(1s^2)$	$(2s^2\ 2p^6)$	$3s^2$
1.00 each	0.85 each	0.35 each
(Rule 5)	(Rule 4)	(Rule 3)

In the third shell (n = 3) only contribution by two electrons in 3s subshell is considered, the electron for which screening constant is calculated is omitted.

∴ Screening constant S = Sum of all the contribution

∴ $S = (2.0 \times 1.0) + (8 \times 0.85) + (2 \times 0.35)$

∴ $S = 2.0 + 6.8 + 0.7 = 9.5$

Thus, $Z^* = Z - S$

$Z^* = 13 - 9.5 = 3.5$

SOLVED EXAMPLES

Example 2.1 : *Calculate Z* for the valence electron of K.*

Solution : Potassium, K (Z = 19), its electronic configuration is,

$K (Z = 19) = 1s^2, 2s^2, 2p^6, 3s^2, 3p^6, 4s^1$

Rule 1, grouping as $(1s^2)$ $(2s^2\ 2p^6)$ $(3s^2\ 3p^6)$ $(4s^1)$

Thus, rule 4, 5, contribution as $(1s^2)$ $(2s^2\ 2p^6)$ $(3s^2\ 3p^6)$

$(1s^2)$ $(2s^2\ 2p^6)$	$(3s^2\ 3p^6)$
1.00 each	0.85 each
Rule 5	Rule 4

∴ $S = (10.0 \times 1.0) + (8 \times 0.85)$

$S = 10.0 + 6.8 = 16.8$

∴ $Z^* = Z - S$

$= 19.0 - 16.8$

$= 2.2$

∴ $Z^* = 2.2$

Example 2.2 : *Calculate the valence electron in sulphur.*

Solution : Sulphur = Z = 16,

Electronic configuration S (Z = 16) = $1s^2, 2s^2, 2p^6, 3s^2, 3p^4$

Rule 1, grouping as $(1s^2)$ $(2s^2\ 2p^6)$ $(3s^2\ 3p^3)$.

The electron whose screening constant is to be calculated (4th electron in 3p subshell) is omitted.

Rule 3, 4, 5, contribution as,

$(1s^2)$	$(2s^2\ 2p^6)$	$(3s^2\ 3p^3)$
1.0 each	0.85 each	0.35 each
Rule 5	Rule 4	Rule 3

$$\therefore \quad S = (2.0 \times 1) + (8 \times 0.85) + (5 \times 0.35)$$
$$2 + 6.8 + 1.75 = 10.55$$
$$\therefore \quad Z^* = Z - S$$
$$= 16 - 10.5 = 5.45$$
$$Z^* = 5.45$$

Example 2.3 : *The 3d electron in chromium.*

Solution : Chromium Z = 24, electronic configuration is,

$$Cr\ (Z = 24) = 1s^2, 2s^2, 2p^6, 3s^2, 3p^6, 3d^5, 4s^1$$

We want to find out the screening constant of any one electron in the 3d subshell.

Rule 1 grouping as $(1s^2)\ (2s^2\ 2p^6)\ (3s^2\ 3p^6)\ (3d^5)\ (4s^1)$

Rule 2, 3, 6, contribution as

$$\underbrace{(1s^2)\quad(2s^2\ 2p^6)\quad(3s^2\ 3p^6)}_{\substack{1.00\text{ each}\\ \text{Rule 6}}}\quad \underset{\substack{0.35\text{ each}\\ \text{Rule 3}}}{(3d^4)}\quad \underset{\substack{\text{zero}\\ \text{Rule 2}}}{(4s^1)}$$

$$\therefore \quad S = (18.0 \times 1.0) + (4 \times 0.35) = 19.4$$
$$Z^* = Z - S$$
$$= 24.0 - 19.4 = 4.6$$
$$Z^* = 4.6$$

Example 2.4 : *Calculate the effective nuclear charge for valence electron of helium.*

Solution : This is unique case where no discrimination between inner electrons and outer electrons can be made. Helium has Z = 2, only two electrons are there in 1s subshell. Only one electron from 1s subshell contributes to screening. This contribution is equal to 0.30 (Rule 3).

$$\therefore \quad S = 0.3$$
$$\text{and} \quad Z^* = 2.0 - 0.3$$
$$Z^* = 1.7$$

Example 2.5 : *Calculate Z* for valence electron of fluorine.*

Solution : Fluorine (Z = 9), electronic configuration is,

$$F\ (Z = 9) = 1s^2, 2s^2, 2p^5$$

Here, we want to find out the screening constant for any one electron from 2p orbital.

Rule 1, grouping as $(1s^2)\ (2s^2\ 2p^4)$

Rule 3, 4, contribution as,

$(1s^2)$	$(2s^2\ 2p^4)$
0.85 each	0.35 each
Rule 4	Rule 3

$$\therefore \quad S = (2.0 \times 0.85) + (6.0 \times 0.35)$$

$$= 1.7 + 2.1 = 3.8$$

$$\therefore \quad Z^* = Z - S$$

$$= 9 - 3.8 = 5.2$$

$$Z^* = 5.2$$

PRACTICE QUESTIONS

(A) Multiple Choice Questions (MCQs) :

Each of the following question is followed by four answers. Select the correct answer.

1. The nucleus of an atom is always ......
   (a) positively charged (b) negatively charged
   (c) neutral (d) none of these
2. The maximum number of electrons in an orbit (shell) or energy level is ......
   (a) n^2 (b) $2n^2$
   (c) $2n$ (d) $2n + l$
3. The total number of orbitals in a shell is ......
   (a) $n - 1$ (b) $n + 1$
   (c) n^2 (d) $2n^2$
4. For a given value of 'n', l has possible values from ......
   (a) $n + 1$ (b) n
   (c) n^2 (d) $n - 1$
5. Capacity of subshell accommodate electrons is given by equation ......
   (a) $2 \times (2l + 1)$ (b) $(2l + 1)$
   (c) $(2l - 1)$ (d) $2(l + 1)$

6. (n – 1) shell is called as ......
   (a) valence shell (b) penultimate shell
   (c) antipenultimate shell (d) none of these
7. Screening constant 'S' and Z* for the valence electron of Na (Z = 11) is ......
   (a) S = 8.8, Z* = 2.2 (b) S = 8.6, Z* = 2.4
   (c) S = 9.0, Z* = 2.0 (d) none of these
8. The element with atomic number 31 is present in which block of periodic table ?
   (a) s-block (b) p-block
   (c) d-block (d) f-block
9. Of all the group-2 elements, the one with the highest ionization energy is ......
   (a) Be (b) Mg
   (c) Ca (d) Sr
10. Correct order of electronegativity of halogen family is ......
    (a) I > Br > Cl > F (b) Cl > Br > I > F
    (c) F > Cl > Br > I (d) none of these
11. According to Aufbau principle, the electrons are added first to orbitals ......
    (a) in sublevels of highest energy
    (b) in sublevels of lowest energy
    (c) in sublevels of zero energy
    (d) none of these
12. The transition elements have a general electronic configuration ......
    (a) $ns^2np^6nd^{1-10}$ (b) $(n-1)d^{1-10}ns^{0-2}np^{0-5}$
    (c) $(n-1)d^{1-10}ns^{0-2}$ (d) $nd^{1-10}ns^{0-2}$
13. The most stable orbital among the following orbitals is ......
    (a) d^1 (b) d^4
    (c) d^5 (d) d^9
14. f-block elements are also called ......
    (a) inner-transition elements(b) lanthanides
    (c) actinides (d) transition elements

(B) State True or False Questions :

1. The arrangement of electrons in various shells, subshells and orbitals of an atom in its ground state is called its electronic configuration.
2. The principal quantum number can have integral values 0, 1, 2, 3 ... etc.
3. The angular momentum (azimuthal) quantum number (l) describes the shape of the orbital.
4. For a given value of 'l' there are $(2l + 1)$ integral values of m_l.
5. All s-orbitals are spherical in shape but differ in size which increases with increase of value of principal quantum number 'n'.
6. No two electrons in an atom can have the same set of four quantum numbers.
7. In neutral isolated atom, the orbital with the lower value of $(n + l)$ has highest energy.
8. Cation is smaller than parent atom.
9. Atomic radius is equal to internuclear distance 'd' between two atoms.
10. Ionization energy decreases down the group.
11. The halogens have highest electron affinity values.
12. Transition elements have incompletely filled 'd' subshells in one of the common oxidation state.
13. An electron configuration with minimum spin multiplicity represents ground state of an atom.

(C) Short and Long Answer Type Questions:

1. Define and explain the following terms :
   (a) Valence shell (b) Penultimate shell
   (c) Antipenultimate shell (d) Electronic configuration
2. What are quantum numbers ? Name and explain different types of quantum numbers and their quantitative significance.
3. Discuss in detail Pauli exclusion principle. Explain with the help of this principle, how to determine the maximum number of electrons in any shell ?

4. How does the angular momentum quantum number 'l' is related to properties of orbital ?

(**Ans.** 5.27×10^{-15} m s^{-1})

5. Explain why carbon has two electrons in different p orbitals with parallel spins ?
6. The probability of finding two electrons with the same set of four quantum numbers in an atom is zero. Explain the statement.
7. What do you mean by orbital ? Sketch the shapes of s, p and d orbitals.
8. Discuss with the help of schematic diagram the order of filling of electrons in various orbitals in case of multielectron atom.
9. Write short notes on :
   (a) Aufbau principle
   (b) Hund's rule
   (c) $(n + l)$ rule
   (d) Quantum numbers
10. What is $(n + l)$ rule ? How it is useful in building up of the elements?
11. Explain why 4s orbital is filled first than 3d even though the principal quantum number of 4s is higher than 3d ?
12. Write the ground state electronic configuration of following elements : B (5), Ne (10), Cl (17), V (23), Fe (26), Cu (29), I (53), Mo (42), Kr (36), Au (79).
13. Explain why ground state electronic configuration of Cr (24) and Cu (29) are different from what we might expect ?
14. Explain what is meant by a noble gas core. How it is used in writing electronic configuration of the elements ?
15. Draw the orbital diagram of the atoms with the following electronic configuration :
    (a) $1s^2, 2s^2, 2p^6, 3s^2, 3p^6$
    (b) $1s^2, 2s^2, 2p^6, 3s^2, 3p^4$
    (c) $1s^2\ 2s^2, 2p^6, 3s^2, 3p^6, 4s^1, 3d^5$
16. Write the atomic number of the elements whose outermost electronic configuration are :
    (a) $3d^9$
    (b) $3p^3$
    (c) $2p^6$
    (d) $4f^3$
17. Describe the general layout of modern periodic table.

18. Sketch the outline of modern periodic table. Show in it the following :
    (a) Noble gases (b) Halogens
    (c) Transition elements (d) f-block
19. State and explain Mendeleev's periodic law. Why was it corrected ? What is modern periodic law ?
20. Define the following terms :
    (a) Effective nuclear charge (b) Screening constant
    (c) Ionization energy (d) Electron affinity
    (e) Electronegativity
21. Write the outer electronic configuration of :
    (a) The alkali metals (b) The alkaline earth metals
    (c) Halogens (d) Noble gases
    (e) Lanthanides (f) Actinides
22. What are the different types of elements in the periodic table ?
23. How does radius change as we move
    (a) from left to right across the period and
    (b) from top to bottom in a group ?
24. What is ionic radius ? How does the size change when an atom is converted to (a) an anion and (b) a cation ?
25. What are the factors that determine atomic and ionic size ?
26. What is ionization energy ? Explain the factors which influence ionization potential. How does the ionization energy change as we move from
    (a) left to right and (b) top to bottom in a group
27. Second and third ionization energies are higher than first ionization energy. Explain why.
28. Ionization energy of Boron is unexpectedly less than Beryllium. Explain.
29. Explain the trend of ionization energy in a group with suitable example.
30. What is electronegativity ? How is it related to sizes of atoms and their electronic configuration ?

31. Electronegativity is found to increase across a period and it decreases down a group. Explain.
32. Explain the following :
    (i) Ionic radii of alkaline earth metals are smaller than those of corresponding alkali metal ions.
    (ii) The atomic size of transition elements decreases only slightly across the period.
    (iii) Cation is smaller than parent atom and anion is larger than its parent atom.
33. Define the term shielding constant. How it affects the atomic radius of an element ?
34. With the help of Slater's rule, calculate screening constant 'S' and Z_{eff} (Z^*) for the following electrons.
    (a) The valence electron of Na (Z = 11). (**Ans.** S = 8.8, Z^* = 2.2)
    (b) The valence electron of Al (Z = 13). (**Ans.** S = 9.5, Z^* = 3.5)
    (c) 3d electron of Mn (Z = 25). (**Ans.** S = 19.4, Z^* = 5.6)
    (d) 3d electron of Co (Z = 27). (**Ans.** S = 20.1, Z^* = 6.9)
    (e) Valence electron of Ne (Z = 10) (**Ans.** S = 4.15, Z^* = 5.85).

UNIVERSITY QUESTIONS

April 2015

1. Explain periodicity in properties of alkaline earth metals with respect to :
   (i) atomic and ionic size, (ii) oxidation state.

October 2015

1. Write the names and electronic configuration of group III A elements and discuss the trends in atomic size and electronegativity.

April 2016

1. Write the names and electronic configuration of group III A elements.
2. Explain periodicity in properties of alkali metals with respect to ionization energy and oxidation states.

October 2016

1. Define : (i) Ionization energy, (ii) Electronegativity.
2. Explain the periodicity in properties of alkaline earth metals with respect to atomic size and ionization energy.
3. Give the names of group I A elements and write electronic configuration for group I A elements.
4. Explain the periodicity in properties of group V A elements with respect to ionization energy and electronegativity.

April 2017

1. Alkali metals do not show +2 oxidation state. Explain.
2. Write the electronic configuration of Cr (Z = 24) and Cu (Z = 29).
3. What are the similarities of hydrogen with alkali metals ?
4. Give the names and write electronic configuration of group I A elements.
5. Explain the family relationship of alkaline earth metals with reference to :

   (i) Size of atoms and ions, (ii) Ionization energy.

October 2017

1. Explain periodicity in properties of alkali metals with respect to atomic size and oxidation state.
2. Give the names and write electronic configuration of group II A elements.
3. Explain periodicity in properties of group VI A elements with respect to ionization energy and electronegativity.

April 2018

1. Give the statements of Hund's rule and Pauli's exclusion principle.
2. Define interhalogen compounds with suitable examples.
3. Give the names, symbol, atomic number and electronic configuration of Group II A elements.
4. What are the similarities of hydrogen with halogen elements ?

***

Chapter 3...

Chemical Bonding

Contents ...

Linus Carl Pauling (February 28, 1901 - August, 19. 1994) was an American chemist, biochemist, chemical engineer, peace activist, author, educator, and husband of American human rights activist Ava Helen Pauling. For his scientific work, Pauling was awarded the Nobel Prize in Chemistry in 1954. For his peace activism, he was awarded the Nobel Peace Prize in 1962.

Linus Pauling in 1962

He is one of four individuals to have won more than one Nobel Prize (the others being Marie Curie, John Bardeen and Frederick Sanger). Of these, he is the only person to have been awarded two unshared Nobel Prizes, and one of two people to be awarded Nobel Prizes in different fields, the other being Marie Curie.

Pauling was one of the founders of the fields of quantum chemistry and molecular biology. His contributions to the theory of the chemical bond include the concept of orbital hybridisation and the first accurate scale of electronegativities of the elements. Pauling also worked on the structures of biological molecules, and showed the importance of the alpha helix and beta sheet in protein secondary structure. Pauling's approach combined methods and results from X-ray crystallography, molecular model building, and quantum chemistry.

Peter Joseph William Debye (March 24, 1884 - November 2, 1966) was a Dutch-American physicist and physical chemist and Nobel laureate in Chemistry.

Peter Debye

3.1 INTRODUCTION

- We know that, the material around us is made of molecules. We also know that, the molecules are made of atoms. The question then arise in the mind of chemist is why and how the atoms are held together to form the basic unit, molecule? The study of material shows that a variety of attractive forces hold the atoms together in a molecule.
- Elements are composed of extremely small particles, called atoms. Atoms are made of protons, neutrons and electrons. Some substances do not exist in atomic form, for example, hydrogen gas. It is pure element that is made up of two hydrogen atoms known as molecule. A molecule therefore is an aggregate of atleast two atoms held together by chemical forces called a bond. When a molecule is formed from atoms, they undergo rearrangement of electrons in their outermost orbit. Hence the nature of bond holding the atoms together is purely a chemical concept.
- A chemical bond therefore holds the atoms together. Thus, a chemical bond or chemical linkage is defined as the net attractive force which holds atoms, ions or molecules together. In this topic, we have to study and understand the concept of chemical bonding.

3.2 ATTAINMENT OF A STABLE CONFIGURATION

- In the previous section we have defined, what is a molecule. But, however, two questions arise as how do atoms combine to form molecules and why do atoms form bonds ? A molecule will only be formed if it is more stable, and has a lower energy, than the individual atoms.
- To understand what is happening in terms of electronic structure, let us consider a group of elements called as inert gases or noble gases, which are well known for their chemical inertness. Atoms of the noble gases do not normally react with any other atoms, and their molecules are monoatomic, i.e. contain only one atom. The very little tendency to react is mainly because the atoms already have a low energy, and it cannot be lowered further by forming the compound. The low energy of these elements is due to their completely filled outer shell by electrons. This is called as a **noble gas structure**, and it is an unusually stable arrangement of electrons.
- Normally, atoms of various elements combine with one another by undergoing electronic rearrangement so as to acquire stable noble gas structure of eight electrons. This simple explanation was independently given by Lewis and Kossel in 1916. It is the electronic concept of chemical bonding called as **Octet Theory**.

- The noble gas have an outer electronic configuration of **ns^2 np^6** (except He, whose electronic configuration is $1s^2$). Thus, it was concluded that the electronic configuration **ns^2, np^6** should possess maximum stability and minimum energy.
- Atoms of all other elements tend to acquire this configuration (i.e. ns^2, np^6) of maximum stability and minimum energy.

3.3 TYPES OF CHEMICAL BONDS

- Atoms can attain a stable electronic configuration in three different ways viz. by losing electrons, by gaining electrons, or by sharing electrons.
- Elements may be divided into three different groups as:
  - (i) Electropositive elements, whose atoms give up one or more electrons fairly readily.
  - (ii) Electronegative elements, which accept electrons fairly easily and
  - (iii) Elements which have little tendency to lose or gain electrons.
- Depending on this fact three different types of bonds can be formed as:

Electropositive element + Electronegative element $\Rightarrow$ Ionic bond

Electronegative element + Electronegative element $\Rightarrow$ Covalent bond

Electropositive element + Electropositive element $\Rightarrow$ Metallic bond

- The above mentioned bonds are idealized or extreme representations. These divisions, however, are made for convenience and most molecules possess bonds of intermediate types. For example, the bond in the lithium chloride is mainly ionic in nature, however, it also possesses a little covalent character. Thus, only few bonds are purely ionic, covalent or metallic.

- Apart from this one more type of bond called as co-ordinate bond exists in some compounds. A brief description of these bonds is given below.

3.3.1 Ionic Bond

- Ionic bonds are formed when electropositive elements react with electronegative elements. In this, complete transfer of one or more electrons from the outermost shell of one atom to the outermost shell of another (dissimilar) atom is done, so that both the atoms achieve a noble gas electronic configuration.
- The atom which loses the electron, forms a positive charge and becomes a positive ion (cation). While, the other atom which gains electron, acquires a negative charge and becomes a negative ion (anion).
- These two oppositely charged ions combine due to electrostatic force of attraction to form a **electrovalent** or **ionic** compound.
- The *electrostatic force of attraction* which holds the oppositely charged ions together and thus results in a decrease in energy of system is known as **ionic or electrovalent bond**.
- Consider, for example, the formation of Lithium fluoride, from lithium and fluorine. The electronic configuration of lithium is $1s^2\ 2s^1$ and that of fluorine is $1s^2\ 2s^2\ 2p^5$. When lithium and fluorine atoms come in contact with each other, the outer $2s^1$ valence electron of lithium is transferred to the fluorine atom. Using Lewis dot symbols, we represent the reaction as:

$$\cdot\text{Li} \quad + \quad \cdot\ddot{\underset{\cdot\cdot}{\text{F}}}: \longrightarrow [\text{Li}]^+ \quad [:\ddot{\underset{\cdot\cdot}{\text{F}}}:]^- \text{ (or LiF)}$$

$$1s^2, 2s^1 \quad 1s^2, 2s^2, 2p^5 \qquad 1s^2 \quad 1s^2, 2s^2, 2p^6$$

- The Li^+F^- unit is held together by the electrostatic attraction between the positively charged lithium and the negatively charged fluoride ion. The result of this attraction is an ionic bond, which holds ions together in an ionic compound. The potential energy (E) is a negative quantity, and the formation of an ionic bond from Li^+ and F^- is an exothermic process which favours a stable bonded pair of Li^+ and F^- ions.
- There are many other common reactions that lead to the formation of ionic bonds. For instance, calcium burns in oxygen to form calcium oxide as:

$$2\ Ca\ (s) + O_2\ (g) \longrightarrow 2\ CaO\ (s)$$

In Lewis dot symbols,

$$\cdot Ca \cdot \quad + \quad \ddot{\underset{\cdot\cdot}{O}}: \longrightarrow [Ca]^{2+} \quad [:\ddot{\underset{\cdot\cdot}{O}}:]^{2-}$$

$[Ar]4s^2$ $\quad 1s^2 2s^2 2p^4$ $\quad [Ar]$ $\quad [Ne]$

There is a transfer of two electrons from the calcium atom to the oxygen atom.

Few more examples are:

1. $Na_{(g)} + Cl_{(g)} \longrightarrow NaCl_{(s)}$

   i.e. $Na\cdot \quad + \cdot\ddot{\underset{\cdot\cdot}{Cl}}: \longrightarrow Na^+ \; Cl^-$ i.e. NaCl

   (2, 8, 1) (2, 8, 7) (2, 8) (2, 8, 8)

2. $Mg + Cl_2 \longrightarrow 2MgCl_2$

   i.e. $:\ddot{\underset{\cdot\cdot}{Cl}}\cdot \quad + \quad \ddot{Mg} \quad + \quad \cdot\ddot{\underset{\cdot\cdot}{Cl}}: \longrightarrow Mg^{++} \quad 2Cl^-$ i.e. $MgCl_2$

   (2, 8, 7) (2, 8, 2) (2, 8, 7) (2, 8) (2, 8, 8)

- The number of electrons lost or gained by an atom is called the **electrovalency** of that element. Larger the difference of electronegativity between two atoms, greater will be the possibility of ionic bond formation.

3.3.2 Covalent Bond

- When two electronegative atoms react together, both atoms have a tendency to gain electrons, but neither atom, has any tendency to lose electrons. In such cases, the atoms share electrons so as to attain a noble gas configuration.
- The bond formed by the ***mutual sharing of electron*** **pairs between the atoms of the same or different elements is called a *covalent bond or electron pair bond.*** In each pair of shared electrons, one electron is contributed by each atom.
- Let us consider the fluorine molecule, F_2. The electron configuration of F is $1s^2 \ 2s^2 \ 2p^5$. The 1s electron do not participate in bond formation, since they are low in energy and spend most of the time near the nucleus. Thus each F atom has seven valence electrons (the 2s and 2p electrons). There is only one unpaired electron on F, so that the formation of the F_2 molecule is represented as:

$$:\ddot{\underset{\cdot\cdot}{F}}\cdot \; + \cdot\ddot{\underset{\cdot\cdot}{F}}: \longrightarrow :\ddot{\underset{\cdot\cdot}{F}} - \ddot{\underset{\cdot\cdot}{F}}: \quad \text{or} \quad :\ddot{\underset{\cdot\cdot}{F}}:\ddot{\underset{\cdot\cdot}{F}}:$$

- It is to be noted that same valence electrons are not involved in covalent bond formation; these are called non-bonding electrons, or lone pairs. Thus each F in F_2 has three lone pairs of electrons.
- In the example of F_2 or Cl_2, each combining atom has only one unpaired electron to share, as a result, only one covalent bond between two atoms can be formed. In different words, we say these covalency in these examples is one or these elements are **univalent**.
- Now consider the oxygen atom, whose electronic configuration is [He] $2s^2$, $2p^4$, which can be shown as

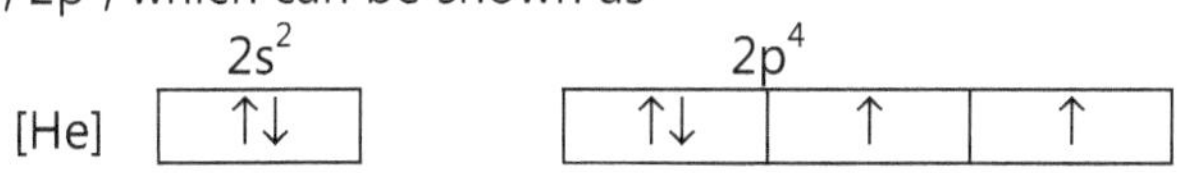

- It may form two covalent bonds by sharing its two unpaired electrons. Therefore an element like oxygen is bi-covalent, as may be seen from the following example:

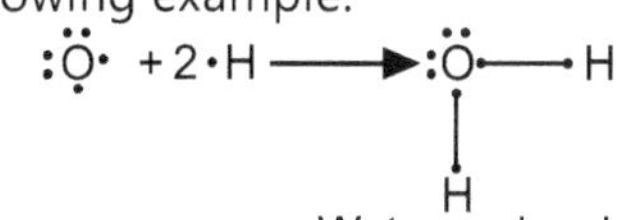

Water molecule

- In a similar way a molecule of ammonia, NH_3 is made up of one nitrogen and three hydrogen atoms as:

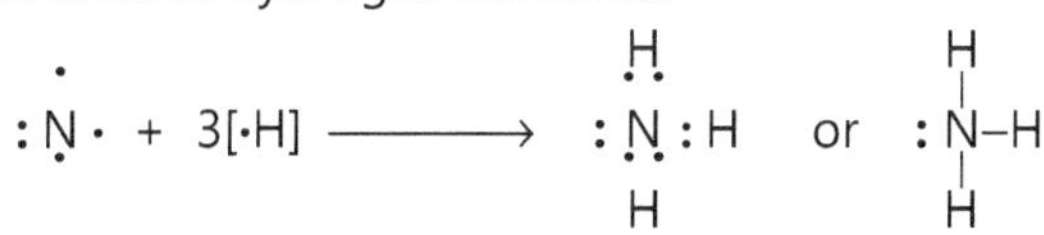

- The nitrogen atom is three electrons short of the noble gas structure, so it forms three bonds, and the hydrogen atoms are one electron short, so they each form one bond. By sharing electrons in this way, both nitrogen and all three hydrogen atoms attain a noble gas structure. Here, the nitrogen is tricovalent.
- Carbon and silicon are tetracovalent in methane and silicon tetrachloride, as shown below:

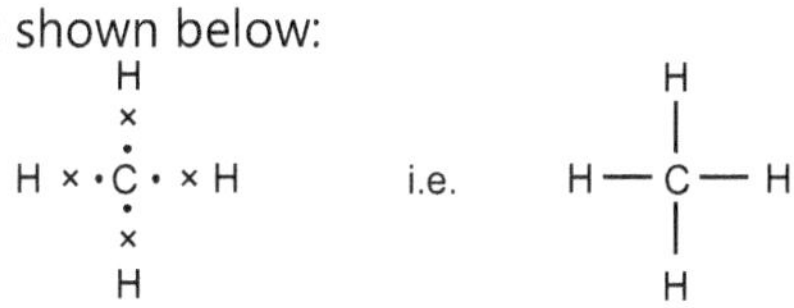

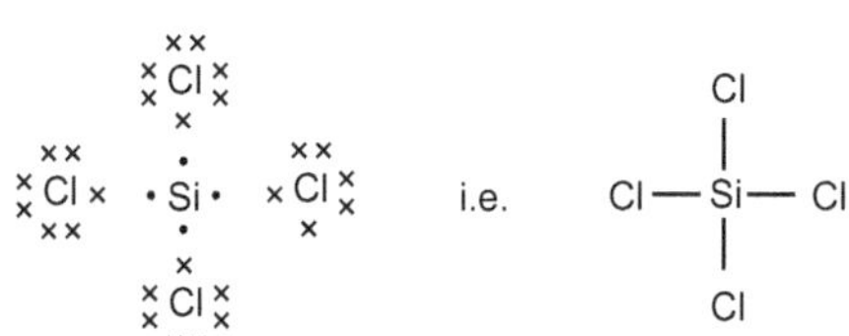

- A covalent bond is directed in space. So the atoms in a covalent compound are linked to each other in definite position. In other words, a covalent compound has definite shape or geometry.

Double and Triple bonds: **(Oct. 2011)**

- Sometimes more than two electrons are shared between a pair of atoms. If four electrons are shared, then there are two bonds, and this arrangement is called a **double bond**. Double bonds are found in molecules like oxygen (O_2), carbon dioxide (CO_2) etc.

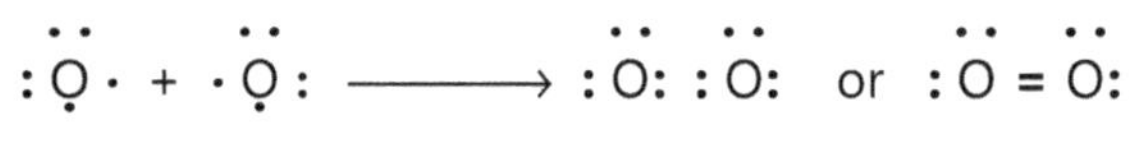

- If six electrons are shared then there are three bonds, and this is called a **triple bond**. Triple bonds are found in molecules like nitrogen (N_2), acetylene etc.

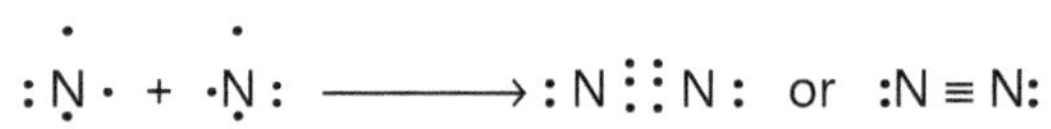

Comparison between Ionic bond and Covalent bond

Ionic bond	Covalent bond
1. It is formed by the transfer of electrons from an electropositive atom to an electronegative atom.	1. It occurs due to sharing of electrons. Shared pair(s) is contributed by both the atoms forming the bond.
2. Ionic bond is weaker.	2. Covalent bond is stronger.
3. Ionic bond has no fixed direction.	3. It has fixed direction.
4. It is formed between atoms of different elements.	4. It may be formed between atoms of same or different elements.
5. It occurs when the difference of electronegativity is large.	5. It occurs when the difference of electronegativity is small.

3.3.3 Co-ordinate Bond **(April 18)**

- It is a special type of covalent bond in which both the electrons required for sharing between two atoms are contributed by one atom only. The atom supplying both the electrons usually possesses one or more lone pair of electrons in its octet and donates it to the other atom which is in need of the two electrons to complete its octet. Thus both the atoms acquire the stable inert gas configuration. The atom which contributes a lone pair of electrons is called as

donor and the atom which is in need of the electrons or which accepts lone pair is called as **acceptor**. The co-ordinate bond is usually denoted by an arrow ($\longrightarrow$). The arrow head indicates the direction in which the donation of electron is made. Co-ordinate bond is also many times called as **co-ordinate covalent bond or dative bond**.

$$\underset{\text{(Donor)}}{X:} + \underset{\text{(Acceptor)}}{Y} \longrightarrow \underset{\text{Co-ordinate bond}}{X: \rightarrow Y}$$

- The most simple and common example of the co-ordinate bond formation is the ammonium NH_4^+ ion. Even though the ammonia molecule has a stable electron configuration, it can react with a hydrogen ion H^+ by donating a share in the lone pair of electrons, forming the ammonium ion as

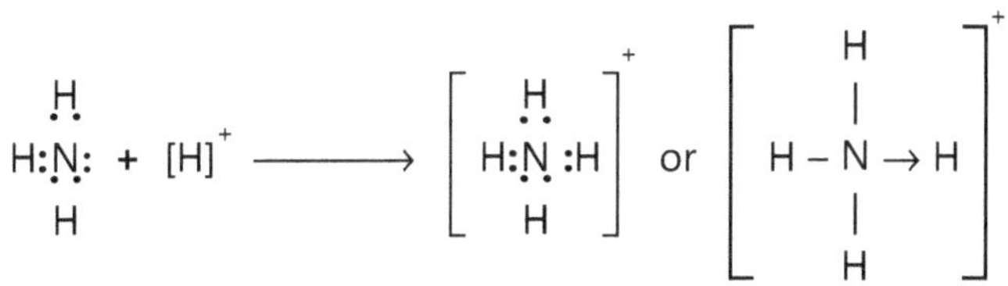

- In a similar way, a molecule of BF_3 can form co-ordinate bond by accepting share in a lone pair from a F^- ion.

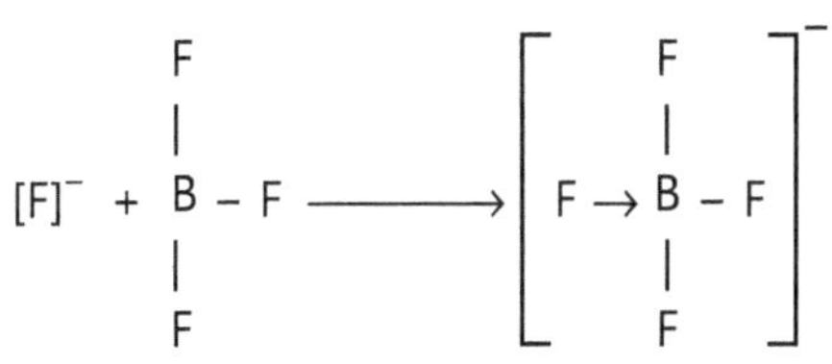

- There are many other examples including:

$$PCl_5 + Cl^- \longrightarrow [PCl_6]^-$$
$$NH_3 + BF_3 \longrightarrow H_3NBF_3$$

- Where co-ordinate bond is present. From all these examples, we can clearly see that co-ordinate bonds differ from normal covalent bonds only in the way they are formed, and once formed they are identical to normal covalent bonds. Thus the mode of formation is different.

3.3.4 Metallic Bond

- The nature of bonding in metals is different from ionic, covalent and co-ordinate type of bonds. The atoms in metals are believed to be held together by means of a special type of bond, known as metallic bond.

- Metals cannot have ionic bond since all the atoms in a metal are identical, while ionic bond requires two atoms with a large difference in electronegativity. Similarly, metals cannot have covalent or co-ordinate bond because a metal atom is surrounded by and joined to large number of similar atoms like eight to twelve atoms, also a metal atom has less number of valence electrons like 1, 2 or 3 and so it cannot form 8 to 12 covalent bonds with the neighbouring atoms.
- In a metal the bonding electrons are delocalized over the entire crystal. In fact, metal atoms in a crystal can be imagined as an array of positive ions immersed in a sea of delocalized valence electrons (Refer Fig. 3.1). The great cohesive force resulting from delocalization is responsible for the strength of metals (metallic bonds). The mobility of the delocalized electrons makes metals good conductors of heat and electricity.

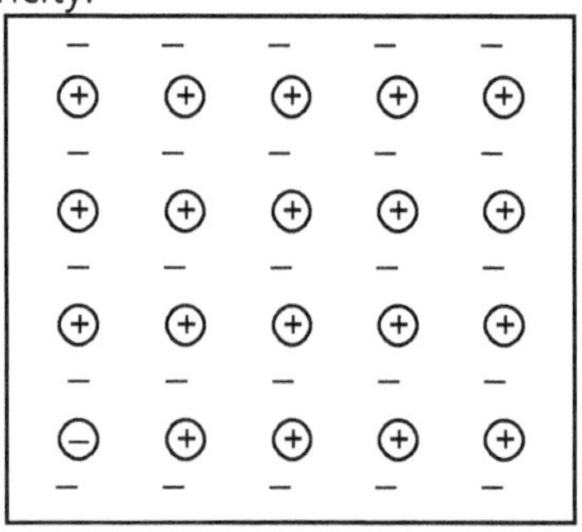

Fig. 3.1: A cross-section of the crystal structure of a metal

3.4 IONIC BOND

- According to the octet theory, formation of ionic bond arises from the loss and gain of electrons by atoms of the elements to attain electronic configuration of the inert gas atoms. By loss and gain of electrons, respectively cations and anions are obtained. Ionic solids contain cations and anions. These ions are held together by ionic bond. For example, in the formation of NaCl, positively charged Na^+ ion is obtained by loss of one electron from neutral sodium atom and negatively charged Cl^- ion is obtained by gain of one electron by neutral chlorine atom.

$$Na_{(g)} \rightarrow Na^+_{(g)} + e^-$$

$$Cl_{(g)} + e^- \rightarrow Cl^-_{(g)}$$

$$Na^+_{(g)} + Cl^-_{(g)} \rightarrow NaCl_{(s)}$$

- The electrostatic force of attraction which keeps Na^+ and Cl^- ions together is known as ionic bond.

Definition :

- **The interionic electrostatic attraction between the oppositely charged ions is known as the ionic bond.**
- The ionic bonds are formed largely between the cations of strongly electropositive elements and the anions of strongly electronegative elements. Ionic bond is non-directional. Whether an ionic compound is likely to be formed or not is decided by :
  (a) ionisation potential,
  (b) electron affinity and
  (c) lattice energy.
  A stable ionic compound is favored by
  (i) Low ionisation potential of the metal.
  (ii) High electron affinity with other elements.
  (iii) High lattice energy of the resulting compound.
  Other ionic compounds are K_2SO_4, CaF_2, $BaSO_4$, etc.

3.4.1 Properties of Ionic Substances

Simple properties of ionic compounds are:

1. Ionic compounds have very low or zero electrical conductivities as solids, but conduct electricity quite well when molten.
2. Ionic compounds tend to have high melting points.
3. Ionic compounds usually are very hard, but brittle substances. The hardness increases with increasing ionic charge and decreasing inter ionic distances.
4. The ionic compounds are often soluble in polar solvents with high dielectric constants, but are insoluble in non-polar solvents such as CCl_4, C_6H_6, etc.

3.4.2 Types of Ions

- An ion is an atom or molecule that has a net electrical charge. Since the charge of the electron is conventionally taken as negative and is equal and opposite to that of proton (considered positive), the net charge of an ion is therefore non zero because of unequal number of protons and electrons. A positive ion called cation is therefore a species which contains less electrons than protons. While an anion is negatively charged with more electrons than protons.
- Thus there are two types of ions, **Positively charged ion is called cation** and **negatively charged ion is called anion**.
- Ion containing only one atom is called monoatomic ion while two or more atoms form polyatomic or molecular ion.

3.4.3 Energy Considerations in Ionic Bonding

- Ionic bonding is a type of chemical bonding that arises from mutual attraction between oppositely charged ions. Therefore ions normally do not exist at their own but combine with each other to form ionic compound (crystal lattice). During the process of formation of ionic compound, certain amount of energy is released called lattice energy or crystal energy.

3.4.4 Lattice Energy (Crystal Energy)

- Lattice energy or crystal energy can be defined as follows :
  1. Lattice energy of an ionic crystal is the energy released when appropriate number of gaseous cations and gaseous anions are brought together from infinity to form one gram mole of the ionic solid at 0°K.

     i.e. $Na^+ (g) + Cl^- (g) \rightleftharpoons NaCl (s), \Delta H = -782 \text{ kJ mol}^{-1}$

     During the formation of a crystal from gaseous ions, energy is released. ΔH is having negative sign.

     Lattice energy of the ionic solid cannot be measured directly, but it is related to other thermodynamic quantities which are measurable.
  2. Alternatively, **lattice energy is defined as the *energy required to convert one gram mole of ionic solid into gaseous ions and bring them to rest at infinity at 0 °K.***

     i.e. $NaCl (s) \rightarrow Na^+ (g) + Cl^- (g), \quad \Delta H = +782 \text{ kJ mol}^{-1}$

     Lattice energy is expressed in terms of unit of heat. It is generally designated as **'U'**. The stability of the ionic solid depends on the value of lattice energy. **Greater the value of lattice energy, more is the stability of the crystal.**

3.4.5 Born-Lande Equation

Theoretically, lattice energy is given by **Born-Lande** equation as below.

$$U = -\frac{NAe^2 (Z^+)(Z^-)}{r}\left(1 - \frac{1}{n}\right)$$

where
U = Lattice energy of the ionic crystal
N = Avogadro's number, $6.023 \times 10^{23} \text{ mol}^{-1}$
A = Madelung constant whose value depends on the geometry of the crystal
Z^+ and Z^- = Charges on cations and anions respectively

r = Internuclear distance between cation and anion

n = Born exponent (associated with repulsion between ions)

From Born-Lande equation, we observe that

(i) **Lattice energy is directly proportional to the product of the charges on the ions.**

For example, $Li^{+} F^{-} \rightarrow -1034$ kJ mol^{-1}

$Ca^{++} 2F^{-} \rightarrow -2606$ kJ mol^{-1}

(ii) **Lattice energy is inversely proportional to the internuclear distance.**

For example, Crystal	→	LiF	LiCl	LiBr	LiI
U (kJ mol^{-1})	→	– 1034	– 840.1	– 781.2	– 718.4

(iii) **Lattice energy is dependent on the charge of the ion.**

For example, lattice energy for MgO is higher than that of LiF even though ionic radii of corresponding ions are almost similar.

(Li^{+} = 0.65 A°, Mg^{++} = 0.65 A°, O^{--} = 1.40 A°, F^{-} = 1.36 A°)

3.4.6 Solvation Energy

The energy associated with dissolution of solute (solid) in solvent (liquid) is called solvation energy. It may be defined *as the **amount of energy released when one mole of ions undergo solvation***. The energy released due to bonding of solvent molecule and molecule or ion of soluble substance is known as solvation energy. If the energy released is more, solute is more soluble in solvent.

Solvation describes the interaction of solvent with dissolved molecules or ions. In the process of salvation, ions or molecules are surrounded by a shell of solvent. Solubility of solute depends on a competition between lattice energy and solvation energy.

The solvation process will be thermodynamically favored only if the net Gibbs energy of the solution is decreased compared to the Gibbs energy of the separated solvent and solute. Strong solvent – solute interaction make the process of solvation more favorable.

3.5 BORN-HABER CYCLE

- Born-Haber cycle is a thermodynamic cycle which relates lattice energy to other thermodynamic quantities, which are experimentally measured such as heat of vaporization, ionisation potential, electron affinity, etc. The various steps in the formation of ionic solid can be shown graphically as a cycle and hence the name is given. Lattice energy of an ionic solid can be measured experimentally by using Born-Haber cycle.

Determination of lattice energy using Born-Haber cycle :

Consider the formation of solid sodium chloride from sodium metal and gaseous chlorine.

Formation of NaCl (s) can be considered by two different methods.

(I) Direct method :

$$\underset{1\ \text{mole}}{Na\ (s)} + \underset{\frac{1}{2}\ \text{mole}}{\frac{1}{2}Cl_2\ (g)} \rightarrow \underset{1\ \text{mole}}{NaCl\ (s)}, \quad \Delta H_f = -410.9\ \text{kJ mol}^{-1}$$

In direct method, sodium metal combines with gaseous molecular chlorine to form solid sodium chloride as shown above. In this reaction, 1 mole of sodium metal combines with half mole of molecular chlorine gas to form one mole of crystalline sodium chloride. During the formation of sodium chloride, energy released is 411 kJ mol^{-1}. This is called as heat of formation of sodium chloride and is represented by ΔH_f.

(II) Indirect method :

Formation of NaCl(s) from sodium metal and gaseous molecular chlorine can be considered to take place in various stages as follows :

(a) Sublimation of sodium metal :

$$Na\ (s) \xrightarrow{+S} Na\ (g), \quad S_{Na} = +108.4\ \text{kJ mol}^{-1}$$

Here one mole of sodium metal sublimes to gaseous sodium. Heat energy required for the process is known as **sublimation energy** or heat of sublimation and is denoted by S.

(b) Dissociation of molecular chlorine to atomic chlorine :

$$\frac{1}{2}\ Cl_2\ (g) \xrightarrow{+\frac{1}{2}D} Cl\ (g), \quad D_{1/2} = +120.9\ \text{kJ}$$

In this step, half mole of chlorine molecule is dissociated to give one mole of chlorine atom. Heat energy absorbed in the process is known as **heat of dissociation.**

(c) Ionisation of gaseous sodium atoms :

$$Na\ (g) \xrightarrow{+I} Na^+\ (g) + e^-, \quad I_{Na} = +495.4\ \text{kJ mol}^{-1}$$

Here one mole of gaseous sodium atoms are converted to sodium ions by removal of valence electron. In this process, heat energy is absorbed and is known as **ionisation energy or ionisation potential.**

(d) Formation of chloride ions :

$$Cl\ (g) + e^- \xrightarrow{-E} Cl^-\ (g), \quad E_{Cl} = -348.5\ \text{kJ mol}^{-1}$$

In this step, electrons given by sodium atoms are taken by gaseous chlorine atoms forming chloride ions. In this process, energy is released and it is known as **electron affinity of chlorine.**

(e) Formation of solid sodium chloride :

$$Na^+ (g) + Cl^- (g) \xrightarrow{-U} NaCl (s), \quad U = -336.8 \text{ kJ mol}^{-1}$$

- One mole of gaseous Na^+ and one mole of gaseous Cl^- combine together to give one mole of crystalline sodium chloride. Energy is released in this process and by definition it is known as lattice energy, U. The value of lattice energy can be determined from all other energy values which can be measured directly.

Graphically Born-Haber cycle can be represented as shown in Fig. 3.2.

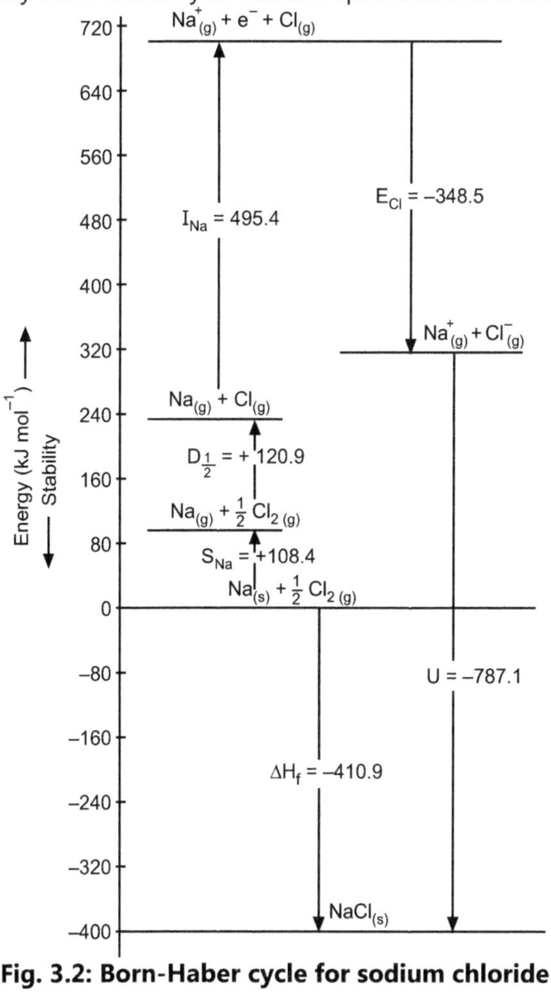

Fig. 3.2: Born-Haber cycle for sodium chloride

- According to Hess's law of constant heat summation, the enthalpy of reaction is the same whether the reaction takes place in one or several steps.

$$\therefore \quad \Delta H_f = S + \frac{1}{2} D + I + E + U$$

- Except lattice energy, all other quantities are experimentally measured and thus lattice energy can be practically determined using Born-Haber cycle.

$$\Delta H_f = S + \frac{1}{2} D + I + E + U$$

$$-410.9 = 108.4 + 120.9 + 495.4 - 348.5 + U$$

$$\therefore \quad U = -410.9 - 108.4 - 120.9 - 495.4 + 348.5$$

$$= -1135.6 + 348.5$$

$$= -787.1$$

Thus, lattice energy of sodium chloride is – 787.1 kJ mol^{-1}.

3.6 APPLICATIONS OF BORN-HABER CYCLE

1. It is difficult to measure electron affinity values of some elements. Therefore, electron affinity is often determined by the relation

$$\mathbf{\Delta H_f = S + D + I + E + U}$$

Under such conditions, lattice energy is calculated using Born-Lande equation.

2. Born-Haber cycle can be used to explain why most of the metals fail to form stable ionic compounds in low valency state. For example, Al, Ca, Mn etc.
3. Born-Haber cycle is useful to know the nature of bonding (ionic or covalent) in the compound. If there is variation in the observed and calculated values of lattice energy, then bond has some covalent character. Greater the deviation from theoretical value of lattice energy, greater is the deviation from ionic character.
4. It can be used as a check on calculated energies.
5. Born-Haber cycle is useful as a means of analyzing and correlating the variations in stability of various ionic compounds.

Example 3.3 : *Calculate the lattice energy of NaI from the following data :*

$$\Delta H_f \text{ of NaI} = -287.6 \text{ kJ mol}^{-1}$$

$$S_{Na} = +108.4 \text{ kJ mol}^{-1}$$
$$I_{Na} = +495.4 \text{ kJ mol}^{-1}$$
$$\frac{1}{2} D_{I_2} = +106.6 \text{ kJ mol}^{-1}$$
$$E = -305.9 \text{ kJ mol}^{-1}$$

Solution :

$$\Delta H_f = S + \frac{1}{2} D + I + E + U$$
$$-287.6 = 108.4 + 106.6 + 495.4 - 305.9 + U$$
$$-287.6 = 710.4 - 305.9 + U$$
$$\therefore \quad U = -287.6 - 710.4 + 305.9$$
$$= \mathbf{-692.1} \quad \text{... Ans.}$$

Thus, lattice energy of NaI is – 692.1 kJ mol^{-1}.

- Experimental values of lattice energy obtained by using Born-Haber cycle and calculated values of lattice energy obtained by using Born-Lande's equation are little different from each other. Experimental values are little higher than calculated values. Table 3.1 lists some common salts and their lattice energy values.

Table 3.1 : Experimental and Calculated lattice energies (–U) of some Alkali Halides (kJ mol^{-1})

Salt	Experimental (Born-Haber cycle)	Calculated (Born-Lande equation)
LiF	1034	1008
LiCl	840.1	811.3
LiBr	781.2	766.1
LiI	718.4	708.4
NaF	914.2	902.0
NaCl	787.1	755.2
NaBr	728.4	718.8
NaI	692.1	663.2

3.7 POLARIZING POWER AND POLARIZABILITY

The ability of a cation to distort anion is known as its polarization power and the tendancy of the anion to become polarized by the cation is known as its polarizability. When two oppositely charged ions come very close to each other, positively charged cation attracts the outermost electrons of the anion. Since some of the electrons of the anion come in between the two ions and ionic bond gets some character of covalent bond. The greater the distortion of charge on the anion, the larger is the tendency of sharing and some covalent character is introduced in the molecule.

The polarizing power and polarizability enhance the formation of covalent bond. This can be explained by studing Fajan's rule.

Fajan's Rule: Fajan's rule explains the cause of polarization of anions. According to Fajan's rule, polarization of anion is favored by:

(a) The charge on ions: Greater the charge on the cation, larger is its polarizing power.

(b) Size of ions: Smaller the cation greater is electrostatic force with which its nucleus will attract electrons of the anion. Due to small size of cation the charge density on the cation is large which increases its polarizing power. Similarly if the size of anion is large, it is easy to polarize an anion because outermost electrons are away from the nucleus and are loosely held.

Fajan's rule can be summarized as below. The formation of bond will be favored by:

Table 3.2

Ionic Bond	Covalent Bond
Low positive charge	High positive charge
Large cation	Small cation
Small anion	Large anion

3.8 BOND MOMENT AND DIPOLE MOMENT

- Some molecules are made of two dissimilar atoms whose electronegativities are far different. In such molecules, for example HF, there is a shift in electron density from H to F because the F atom is more electronegative than the H atom. As a result of this there is a charge separation in the molecule as shown below,

$$\overset{\delta+}{H} - \overset{\delta-}{F}$$

- Slight negative charge is developed on electronegative F atom and slight positive charge is developed on H atom. Such molecules are called polar molecules. This charge separation can be confirmed by keeping the molecule in electric field. When the field is turned on, HF molecule orient itself so that its negative end (pole) is towards the positive plate and their positive end is towards the negative plate.
- A quantitative measure of the polarity of a bond is its dipole moment, μ, which is the product of the charge Q and distance r between the charges.

$$\mu = Q \times r$$

- To maintain electrical neutrality, the charges on both ends must be equal in magnitude and opposite in sign. The quantity Q in equation refers to the magnitude and not sign, so μ that is dipole moment is always positive. Dipole moment is expressed in debye units, D (1 D = 3.33×10^{-30} Cm) where C is coulomb and m is meter.
- Diatomic molecules containing atoms of same elements, for example, H and H in H_2, O and O in O_2, F and F in F_2 do not have dipole moment and so are non-polar molecules. On the other hand, diatomic molecules containing atoms of different elements, for example, H and Cl in HCl, C and O in CO , N and O in NO have dipole moment. The dipole moment of molecule containing three or more atoms, however, depends on both polarity of bonds and geometry of the molecule. Sometimes even if polar bonds are present the molecule may not have a dipole moment. For example, carbon dioxide molecule is triatomic and its geometry is either linear or bent as shown below.

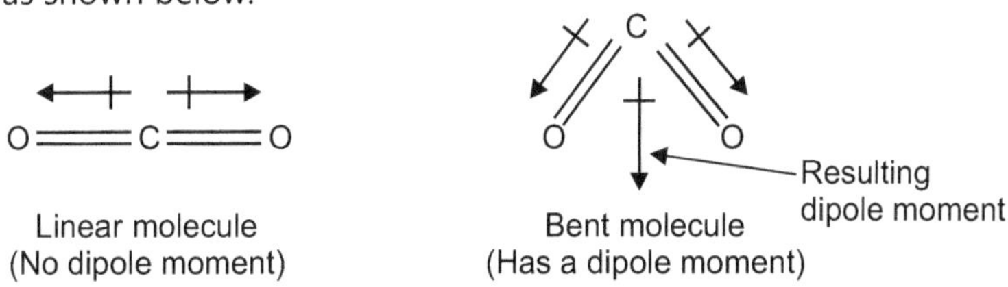

Fig. 3.3

- The arrows show the shift of electron density from the less electronegative carbon atom to the more electronegative oxygen atom. In both cases the dipole moment of CO_2 molecule is due to two **bond moments,** that is, individual dipole moments in the polar C=O bonds. The bond moment is a vector quantity, that is, it has both magnitude and direction. The total bond moment is equal to the vector sum of the bond moments. In the case of linear CO_2 molecule two bond moments are equal in magnitude. Since they point in opposite direction the sum of the bond moment is zero and hence dipole moment is also zero.
- On the other hand, if CO_2 molecule is bent, the two bond moments add to each other and molecule would have dipole moment. Experimentally it is found that carbon dioxide has no dipole moment and is a linear molecule.
- Dipole moment measurements can be used to distinguish between molecules. For example, the following two molecules, both have the

same molecular formula ($C_2H_2Cl_2$), the same number and types of bonds, but different molecular structure as shown below:

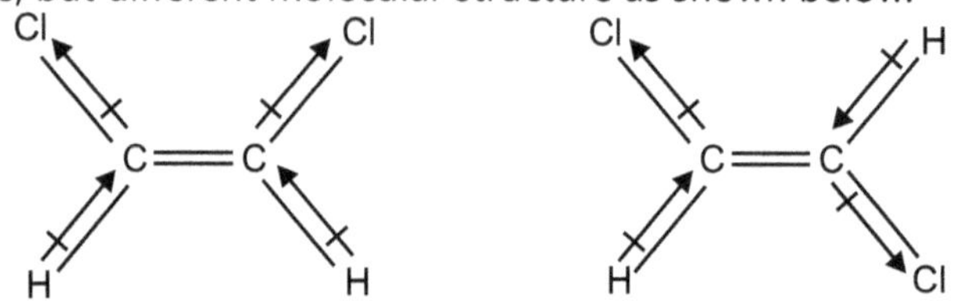

Cis-dichloroethylene
μ = 1.89 D

Trans-dichloroethylene
μ = 0

Fig. 3.4

- Since cis isomer is polar molecule but trans is not, they can readily be distinguished by dipole moment measurement.

3.9 IONIC CHARACTER OF COVALENT BOND

- When two dissimilar atoms are linked by a covalent bond, the shared electron pair is not attracted equally by the two nuclei of bonded atoms. Due to unequal sharing of bonded electron cloud, one end of molecule acquires partial positive charge and the other end acquires partial negative charge (δ+ and δ−). The resulting is thus polar. This polarity of bond gives partial ionic character to the covalent bond.

The percentage of ionic character depends upon the difference of electronegativity of two atoms forming the covalent bond. For example,

(a) HCl molecule: Here electronegativity if hydrogen atom is 2.1 and that of chlorine atom is 3.0. Due to comparatively high electronegativity of chlorine atom, it has ability to pull the bonded pair of electron towards itself and bond between H and Cl is a polar covalent bond. In HCl molecule the covalent bond is 17-18% ionic.

(b) HF molecule: Here electronegativity, if hydrogen atom is 2.1 and that of fluorine atom is 4.0. Due to very high electronegativity of fluorine atom, it has strong ability to pull the bonded pair of electron towards itself and bond between H and F is a polar covalent bond. The difference of electronegativity is very large and in HF molecule the covalent bond is about 64% ionic

3.10 COVALENT BOND

Theories of Bonding:

- Lewis theory of chemical bonding does not clearly explain why chemical bonds exist. Relating the formation of a covalent bond to the pairing of electrons was a step in the right direction, but it did not go far enough. There are certain limitations in this theory, few important ones can be given as

1. It could not explain the cause of covalent bond formation.
2. It could not explain the nature of attractive forces operating between the atoms.
3. It describes single bond between H atoms in H_2 molecule and that between F atoms in F_2 molecule in essentially the same way as the pairing of two electrons. Yet these two molecules have quite different bond dissociation energies and bond lengths (436.4 kJ/mol and 74 picometer for H_2 and 150.62 kJ/mol and 142 picometer for F_2).
4. It could not predict the geometry of covalent molecules. For example it could not explain as why BF_3 is planar and NH_3 is pyramidal.

- Like these and many other facts cannot be explained by Lewis theory. For a more complete explanation of chemical bond formation, we look quantum mechanics. At present, two quantum mechanical theories are used to describe the covalent bond and electronic structure of molecules. Valence bond (VB) theory assumes that the electrons in a molecule occupy atomic orbitals of the individual atoms. The second theory, called molecular orbital (MO) theory, assumes the formation of molecular orbitals from the atomic orbitals. Neither theory perfectly explains all of bonding, but each has contributed something to our understanding. We will discuss valence bond theory here.

Valence Bond Theory

- It is the orbital concept of covalent bond. This concept was first developed by Heitler and London in 1927, to explain the formation of covalent bond in hydrogen molecule.

3.10.1 Heitler-London Theory

The Heitler-London theory is mainly based on following points:

1. A covalent bond is formed when the orbitals of two atoms having unpaired electrons overlap, this is possible only when the two atoms approach within atomic distance.
2. The unpaired electrons in overlapping orbitals must have their spins opposite to each other.
3. During overlapping of two atomic orbitals, the opposite electron spins get neutralised i.e. the electrons get paired leading to the increase in electron density between the two nuclei. This results

in the liberation of energy, increase in attractive forces between the two atoms and thus formation of stable bond between them.

4. If the net result between the two atoms is repulsion, then the total energy of the system increases and bonding does not occur.
5. The equilibrium between the nuclei of the bonded atoms will be that at which attractive and repulsive forces balance each other and the total energy of the combined atom is minimum and hence maximum stability.

3.10.2 Pauling-Slater Theory

Heitler-London theory was extended by Pauling and Slater in 1931 to incorporate the directional characteristics of the bond. The main postulates of the theory are as follows:

1. Overlap takes place between those electron orbitals that are involved in the bond formation. The other orbitals of the atoms do not overlap.
2. Complete overlapping (merging) of two orbitals is not possible due to repulsion between the two nuclei.
3. The overlapping of the orbitals increases the electron density in the region between the two nuclei and each electron is free to move about in the overlapping orbitals, i.e. each electron belongs to both the atomic nuclei.
4. The strength of the covalent bond depends upon the extent of overlapping of the orbitals. The greater the overlapping of orbitals, more is the evolution of energy and stronger is the bond formed.
5. Usually the number of bonds formed is equal to the number of unpaired electrons present in the atom.
6. The atomic orbitals having same symmetry with respect to the molecular axis only overlap to form a bond.
7. In the bond formation, s orbitals being spherically symmetrical can have overlap in any direction The p orbitals can have overlap along x, y or z axis on which the lobes of the p-orbitals lie. Similarly, d and f orbitals are also directed in space and therefore produce overlap in the direction of those orbitals. Covalent bonds are thus considered as directed bonds.

- Thus to summarize our discussion of valence bond theory and to understand it in a better way, let us consider the formation of

H_2 molecule from two H atoms. Lewis theory describes the H-H bond in terms of the pairing of the two electrons on the H atoms. According to valence bond theory, the covalent H-H bond is formed by the overlap of the two 1s orbitals in the H atoms. By overlap, we mean that the two orbitals share a common region in space.

- When the two atoms are far apart, there is no interaction. We say that the potential energy of this system (i.e. the two H atoms) is zero. As the atoms approach each other, each electron is attracted by the nucleus of the other atom. At the same time, the electron repels each other, as do the nuclei.
- The repulsive forces tend to increase the energy of the system, while the attractive forces decrease the energy. Now while atoms are still separated, attraction is stronger than repulsion, so that the net potential energy of the system decreases (i.e. becomes negative) as the atoms approach each other (Refer Fig. 3.5). This trend continues until the potential energy reaches a minimum value. At this point, where the system has the lowest potential energy, it is most stable. This condition corresponds to substantial overlap of the 1s orbitals and formation of a stable H_2 molecule. If the distance between the nuclei were to decrease further, the potential energy would rise steeply and finally become positive as a result of the increased electron-electron and nuclear-nuclear repulsions.

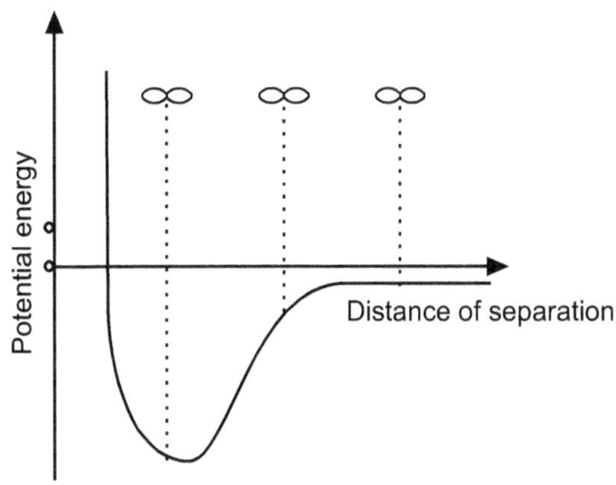

Fig. 3.5: Potential energy diagram for H_2 molecule formation

- In accordance with the law of conservation of energy, the decrease in potential energy as a result of H_2 molecule, formation must be accompanied by a release of energy. Experimentally, it is proved that as a H_2 molecule is formed from two H atoms, heat is given off. The converse is also true. To break H-H bond, some energy must be supplied to the molecule.

- Thus we see that **valence bond (VB) theory** gives a more clear picture of chemical bond formation than the Lewis theory. Valence bond theory states that a stable molecule forms from reacting atoms when the potential energy of the system has decreased to a minimum; the **Lewis theory** ignores energy changes in chemical bond formation.

3.11 CONCEPT OF HYBRIDIZATION

- According to valence bond (VB) theory, chemical bond is formed due to the overlap of atomic orbitals with one another. However, atomic orbital must contain unpaired electron. If we look at the shapes of atomic orbitals, we find that the lobes of p or d orbitals are directed in space in a particular direction. Thus, bond formed by overlapping of such orbitals must be directional and molecule formed must have definite shape. Formation of simple molecules like H_2, F_2, HF etc. can be easily explained on the basis of overlap of pure s and p orbitals. However, this electron pair theory fails to explain the formation of compounds of Be, B, C and many polyatomic molecules. To understand the nature of bond, geometry of the molecule and to explain bonding in molecules, concept of hybridization is found very useful.

3.11.1 Need for the Concept of Hybridization (April 10)

- The electronic configuration of the elements namely Beryllium, Boron and Carbon are given in Table 3.3. The electrons present in 1s orbital are inner electrons and so usually do not take part in bonding. The valence shell of Be, B and C atoms contains 2s and 2p orbitals.

Table 3.3: Electron configuration of some elements

Element	Atomic Number	Electron configuration	Valence shell configuration, box diagram 2s	2p		
Beryllium	4	$1s^2, 2s^2$	↑↓			
Boron	5	$1s^2, 2s^2, 2p^1$	↑↓	↑		
Carbon	6	$1s^2, 2s^2, 2p^2$	↑↓	↑	↑	

- According to electronic configuration and the electron pair theory, Beryllium should not form any covalent bond and must be an inert element, since it has no unpaired electrons in ground state. Boron with one unpaired electron ($2p^1$) should form 1 covalent bond and carbon with two unpaired electrons ($2p^2$) should form two covalent bonds. It is expected therefore, that Beryllium must be zero valent, Boron monovalent and Carbon divalent. But Beryllium forms divalent compounds like BeH_2, BeF_2. Boron forms trivalent compounds like BCl_3, BF_3 and Carbon forms tetravalent compounds like CH_4, CCl_4 etc. Thus, it was not possible to explain the observed valencies of the elements like Be, B and C in their compounds on the basis of the number of unpaired electrons in the ground state of the atom.
- Another important need to introduce the concept of hybridization is to explain the observed geometries of the molecules. The overlapping of atomic orbitals for the formation of covalent bond could explain the geometries of simple molecules like H_2, F_2, HF etc. but, however, it could not explain satisfactorily the shapes of many molecules. For example, V shape of water molecule or tetrahedral shape of the methane molecule. Thus, according to valence bond theory, the formation of water molecule should involve the overlap of the s-orbitals of the two hydrogen atoms with the two p-orbitals of oxygen atom. Since the three p-orbitals of any atom are at right angle to each other, the H–O–H bond angle must be 90°. However, the bond angle is found to be 104° 27' and water molecule is V shaped. Similarly, when carbon combines with four hydrogen atoms to form methane molecule, three of the H–C–H bond angles must be 90° (because p-orbitals are perpendicular to each other) but, the bond angle in all four H–C–H bonds is found to be 109° 28' and methane molecule is tetrahedral in shape.

Thus, there was a need to introduce the concept that explains the correct geometries of the above molecules and many more observations. To explain bonding in methane and many such molecules, valence bond theory uses the concept of hybridization. This concept is hypothetical but is extremely useful.

3.11.2 Definition of Hybridization

- Hybridization can be defined by two different ways as: "*The combining or mixing of the wave functions of the atomic orbitals in linear way to give altogether different wave function of well distributed hybrid orbitals is called as hybridization*".
- The other more simple way to define it is "*the process of mixing and redistribution of atomic orbitals of different energies in an isolated*

atom to form a new set of orbitals of equal energies is called hybridization".

- The new orbitals which are formed by mixing of atomic orbitals are called hybrid orbitals. These hybrid orbitals have a shape of atomic orbitals from which they are formed. The hybrid orbitals have maximum symmetry and definite direction in the shape. For example, the mixing of one s and three p orbitals in this way gives four sp^3 hybrid orbitals. All four orbitals are equal in shape, but different from pure s or pure p orbitals. These four equivalent hybrid orbitals are directed towards the four corners of a regular tetrahedron. The angle between any two orbitals is 109^0 28'.

3.11.3 Important Features of Hybridization

- Hybridization takes place under certain conditions of atomic orbitals and hybrid orbitals formed by mixing of atomic orbitals have certain characteristics. Few important characteristics of hybrid orbitals are given below:
  (i) The orbitals of same atom or ion can undergo hybridization.
  (ii) The orbitals which take part in hybridization should differ only slightly in energy, for example, 2s and 2p orbitals can hybridize, but 2s and 3p it is difficult.
  (iii) The number of hybrid orbitals formed is exactly equal to the number of hybridizing (atomic) orbitals.
  (iv) The shape of hybrid orbitals is different from that of the atomic orbitals from which they are formed. But, however the hybrid orbitals produced are generally of equal energy and identical shape.
  (v) In the hybrid orbitals, the orbital become concentrated in one particular direction so that it is represented as one small lobe and the other large lobe. Due to the concentration in one direction, greater overlapping is possible. Hence hybrid orbitals form stronger bond.
  (vi) Similar to atomic orbitals each hybrid orbital can accommodate maximum two electrons having opposite spins.
  (vii) To avoid mutual repulsion of electrons, the hybrid orbitals orient themselves in space around the central atom in such a way that they are as far as possible away from each other. This gives maximum symmetry and stability to the molecule.
  (viii) The hybrid orbitals can overlap with atomic orbitals or hybrid orbitals of other atoms to form covalent bond.
  (ix) The shape of molecule and the bond angle are suggested by the type of hybridization and orientation of orbitals.

3.11.4 Uses of Hybridization Concept

The concept of hybridization is very useful in solving following problems:

(i) To explain the observed valencies of the elements, for example tetravalency of carbon.
(ii) To predict the geometry and bond angles of molecules.
(iii) To form stronger bond in particular direction and
(iv) To explain equivalent strength of bonds in the molecule.

3.11.5 Steps Involved in Hybridization Process

- The process of hybridization involves following steps:
  (i) Formation of excited state of an atom which involve unpairing of the valence shell electrons.
  (ii) Mixing and recasting of atomic orbitals to produce new equivalent hybrid orbitals.
  (iii) Symmetric arrangement, that is, orientation of hybrid orbitals in space with stable configuration having minimum energy.
  (iv) Overlapping of hybrid orbitals with orbitals of the other atom to form a covalent bond.

3.12 TYPES OF HYBRIDIZATION

- As hybrid orbitals have definite orientations in space, they play a crucial role in determining the shape and geometry of molecules. The type of hybridization of the central atom depends upon nature and number of its orbitals such as s, p, d taking part in process. Accordingly, there are different types of hybridization. To make it simple we divide the type of hybridization into two types as:
  (i) Hybridization involving s and p orbitals such as **sp, sp^2, sp^3** and
  (ii) Hybridization involving s, p and d orbitals such as **dsp^2, sp^3d, sp^3d^2, sp^3d^3** etc.

Hybridization

(A) Using s and p orbitals
(i) sp
(ii) sp^2
(iii) sp^3

(B) Using s, p and d orbitals
(i) dsp^2
(ii) sp^3d
(iii) sp^3d^2
(iv) sp^3d^3

3.12.1 Hybridization using s and p Orbitals

(a) sp Hybridization (Diagonal Hybridization)

- The mixing of one s and one p orbital to form two new equivalent hybrid orbitals of identical energy is known as sp hybridization. Since linear shape of the molecule is obtained in sp hybridization, it is also called as diagonal hybridization. Various examples of the molecules in which central atom shows sp hybridization are BeH_2, BeF_2, C_2H_2, CO_2 etc.

Formation of BeH_2 Molecule:

- In BeH_2 molecule, the central atom is beryllium. Beryllium atom (atomic number = 4) has ground state electronic configuration as $1s^2$, $2s^2$. The valence electron configuration of beryllium atom is $2s^2$. In the ground state, beryllium atom does not contain unpaired electron to form covalent bond with H atoms. But in its excited state, one of the 2s electron is unpaired and then promoted to an empty 2p orbital. Thus, excited state electron configuration becomes $1s^2$, $2s^1$, $2p_x^1$. Now, beryllium atom has two unpaired electrons in 2s and 2p orbitals. One 2s orbital then mixes with one 2p orbital to form a set of two equivalent sp hybrid orbitals. Each sp hybrid orbital contains one unpaired electron, that is they are half filled. This hybridization can be shown with the help of box diagram as below:

		1s	2s	2p		
Beryllium atom (in ground state)	=	↑↓	↑↓			
Beryllium atom (in excited state)	=	↑↓	↑	↑		

		1s	two sp hybrid orbitals		two 2p unhybridized orbitals	
Beryllium atom (in hybridized state)	=	↑↓	↑	↑		

- Due to mutual repulsion of electrons in these orbitals, they orient themselves away from each other as far as possible. This minimizes repulsion between them. Hence they are arranged along a straight line in opposite direction. This makes an angle of 180°. Each hybrid orbital is more concentrated on one side of the nucleus. Thus each sp hybrid orbital consists of one large and one small lobe as shown below (Refer Fig. 3.6).

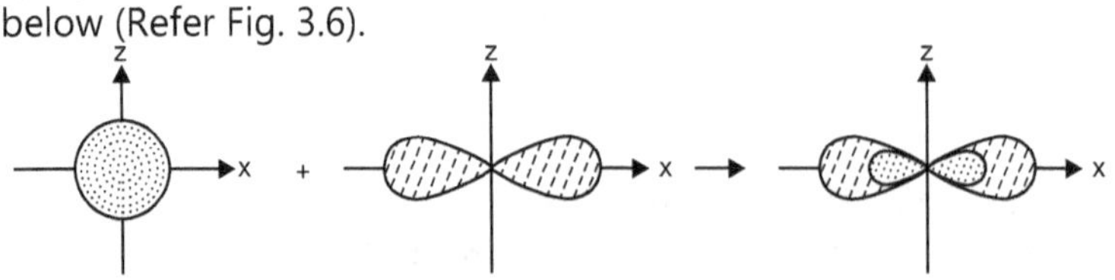

2s orbital **$2p_x$ orbital** **two sp hybrid orbitals**

Fig. 3.6: sp hybridization of Beryllium atom

- The electron configuration of hydrogen atom (atomic number = 1) is $1s^1$, that is, it has one half filled s orbital. Thus, in BeH_2, two s orbitals of two hydrogen atoms overlap axially with sp hybrid orbitals of beryllium atom forming covalent bonds. The overlap is of the type sp-s. Fig. 3.7 shows the formation of BeH_2 molecule.

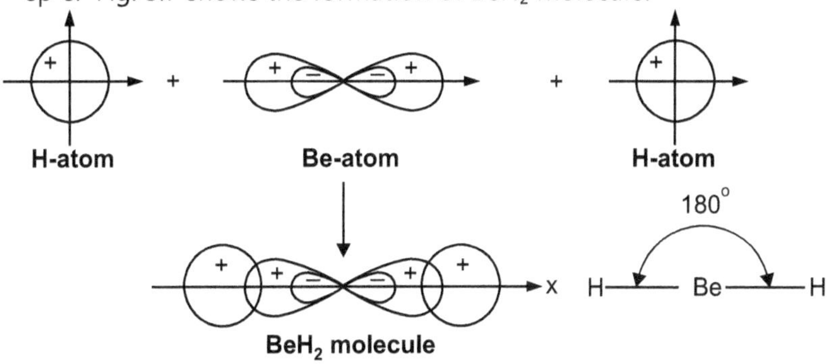

Fig. 3.7: Formation of BeH_2 molecule

- BeH_2 molecule involves two Be-H bonds which are resulted due to overlapping of sp hybrid orbitals on Be atom with that of s orbital of hydrogen atom. BeH_2 molecule is linear and bond angle between H-Be-H is 180°. In this way, co-valency of beryllium in its compounds is 2.

(b) sp^2 Hybridization (Trigonal Hybridization)

- The mixing of one s and two p orbitals to form three new equivalent hybrid orbitals of identical energy and shape is known as sp^2 hybridization. Since trigonal shape of the molecule is obtained after sp^2 hybridization, it is also called as trigonal hybridization. Examples of the molecules in which central atom exhibits sp^2 hybridization are BF_3, BCl_3, etc.
- **Formation of BF_3 molecule:** In BF_3 molecule, central atom is Boron. Its atomic number is 5 and has ground state electron configuration as $1s^2$, $2s^2$, $2p^1$. The valence electron configuration of boron atom is $2s^2$, $2p^1$. Its ground state contains only one unpaired electron, so it cannot form three bonds with fluorine atoms. In its excited state, one of the 2s electron is unpaired and then promoted to an empty 2p orbital. Thus, in the excited state, boron atom possesses three unpaired electrons as $2s^1$, $2p_x^1$ and $2p_y^1$. Now in order to get three equivalent orbitals, boron undergoes hybridization, in which one 2s and 2p orbitals intermixes with each other to give a set of three equivalent sp^2 hybrid orbitals. Each sp^2 hybrid orbital contains one

unpaired electron, that is, they are half filled. This hybridization can be shown with the help of box diagram as below:

Boron atom (in ground state)	=	↑↓ 1s	↑↓ 2s	↑		2p
Boron atom (in excited state)	=	↑↓ 1s	↑ 2s	↑	↑	2p
Boron atom (in hybridized state)	=	↑↓ 1s	↑	↑	↑ three sp^2 hybrid orbitals	↑ one 2p unhybridized orbital

- Due to the mutual repulsion of electrons in these orbitals, they orient themselves away from each other as far as possible. This minimizes repulsion between them. Hence, the three orbitals are coplaner and are directed along three corners of a regular triangle. This arrangement makes an angle of 120^0 with one another. Formation of sp^2 hybrid orbitals is shown below with the help of orbital pictures (Refer Fig. 3.8).

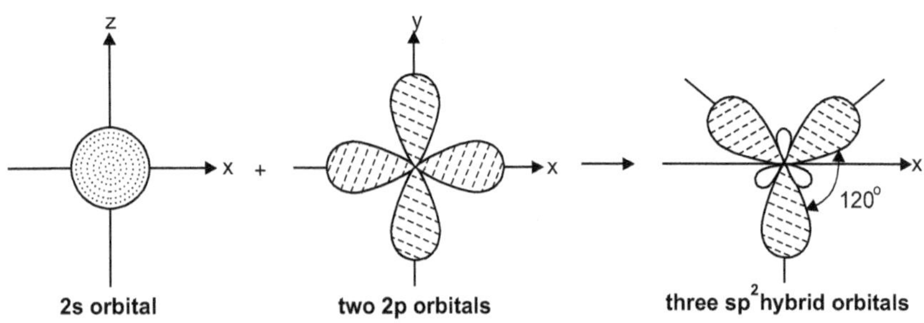

Fig. 3.8: Formation of sp^2 hybrid orbitals

- The electron configuration of fluorine atom (atomic number = 9) is $1s^2$, $2s^2$, $2p^5$, that is, it has one half filled p-orbital. Thus in BF_3 three p orbitals of three fluorine atoms overlap with three sp^2 hybrid orbitals of boron atom forming three covalent bonds in the trigonal plane. The overlap is of the type sp^2-p. Fig. 3.9 shows the formation of BF_3 molecule.

- BF_3 molecule involves three B-F bonds which are due to overlapping of sp^2 hybrid orbitals on B atom with that of p-orbital of fluorine atom. BF_3 molecule is triangular and bond angle between F-B-F is 120^0. Thus, covalency equal to 3, of boron atom in this compound can be explained.

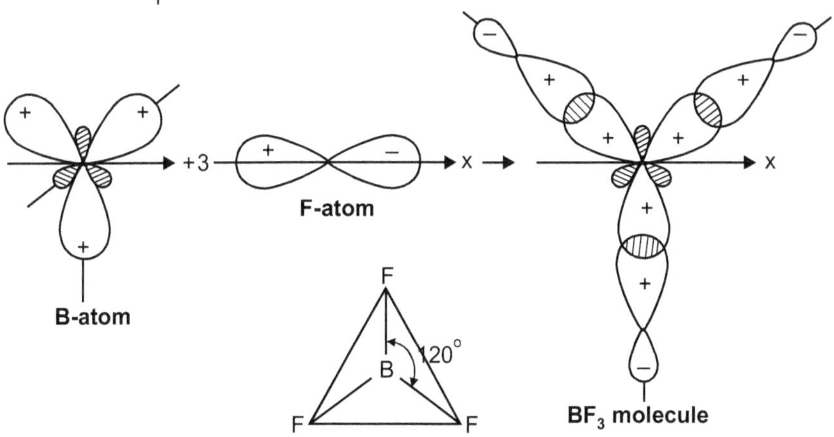

Fig. 3.9: Formation of BF_3 molecule

(c) sp^3 Hybridization (Tetrahedral Hybridization)

- The mixing of one s and three p orbitals to form four new equivalent hybrid orbitals of identical energy and shape is known as sp^3 hybridization. Since tetrahedral shape of the molecule is obtained after sp^3 hybridization, it is also called as tetrahedral hybridization. Examples of the molecules in which central atom exhibits sp^3 hybridization are CH_4, CCl_4, $[Ni(CO)_4]$, $[Mn(Cl)_4]^{2-}$ etc.

Formation of $[Mn(Cl)_4]^{2-}$ complex ion:

- In $[Mn(Cl)_4]^{2-}$ complex ion, central atom is manganese. Its atomic number is 25 and has valence electron configuration as $4s^2$, $3d^5$. In the complex ion, oxidation number of manganese is +2 and therefore its valence electron configuration is $4s^0$, $3d^5$. Thus, manganese ion has half filled 'd' orbital and empty 4s and 4p orbitals. Manganese ion therefore undergoes sp^3 hybridization. One 4s orbital intermix with three 4p orbitals to give a set of four equivalent hybrid orbitals. All four sp^3 hybrid orbitals thus formed are empty and therefore can accept a lone pair of electrons from each chlorine atom. This hybridization can be shown with the help of box diagram as below.

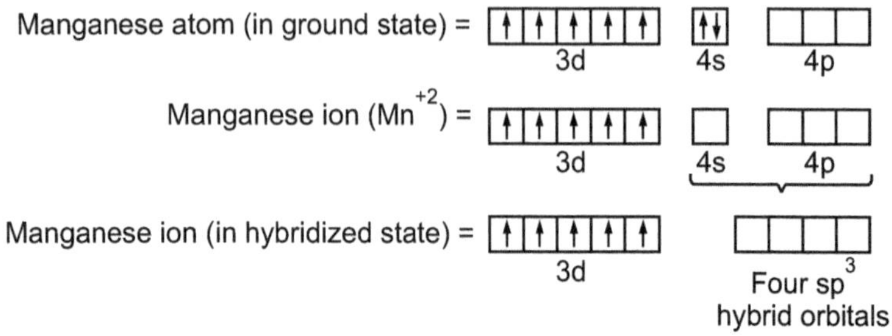

- sp^3 hybrid orbitals orient themselves away from each other as far as possible. The best arrangement in three dimension that keeps them maximum away is of tetrahedron. Thus the four sp^3 hybrid orbitals are directed towards the four corners of a regular tetrahedron making an angle of 109° 28' with one another. Fig. 3.10 shows the formation of sp^3 hybrid orbitals as below:

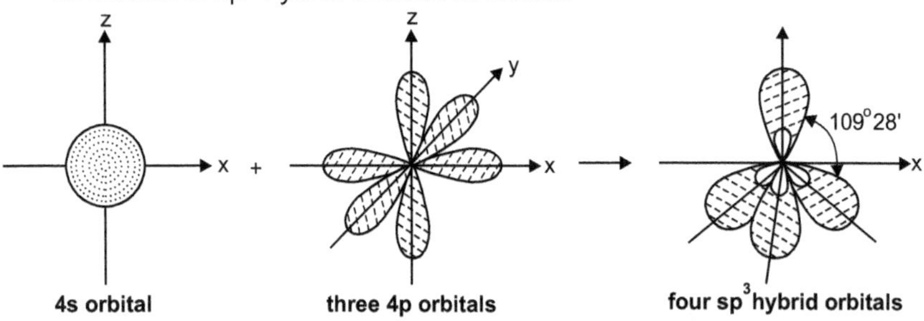

Fig. 3.10: Formation of sp^3 hybrid orbitals

- Chlorine atom has valence electronic configuration as $3s^2$, $3p^5$. It can donate a lone pair of electron into empty sp^3 hybrid orbital on manganese ion to form co-ordinate covalent bond. Following Fig. 3.11 shows the formation of $[Mn(Cl)_4]^{2-}$ ion.

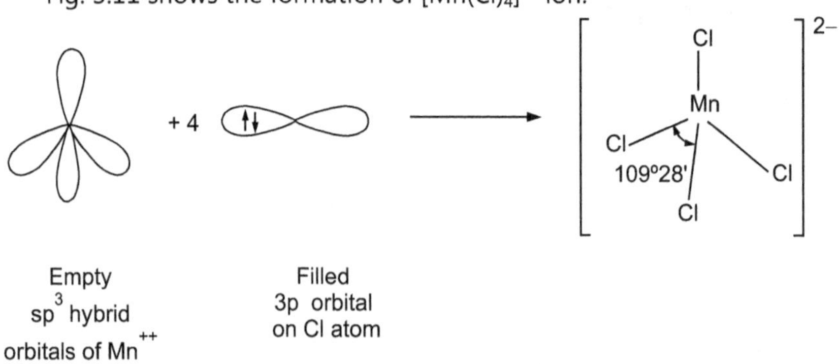

Fig. 3.11: Formation of $[Mn(Cl)_4]^{2-}$ molecule

3.12.2 Hybridization Using s, p and d Orbitals

(a) dsp^2 Hybridization (Square Planar Hybridization):

The mixing of one d, one s and two p orbitals to form four new equivalent hybrid orbitals of equal energy and shape is known as dsp^2 hybridization. Since square planar shape is obtained in this hybridization, this type of hybridization involves the mixing of $d_{x^2-y^2}$ orbital of penultimate shell, while the s and two p orbitals (say p_x and p_y) of the valence shell of central metal atom. The $d_{x^2-y^2}$ orbital lies along x and y axes, p_x orbital lies along x-axis, p_y orbital lies along y-axis, while s-orbital is spherical so after intermixing the four dsp^2 hybrid orbitals obtained are directed to four corners of a square in one particular plane, i.e. xy plane, hence such type of hybridization is also called as square planar hybridization. The examples of the molecules in which central atom exhibits dsp^2 hybridization are $[Ni(CN)_4]^{2-}$, $[Pt(Cl)_4]^{2-}$, $[Pt(NH_3)_2Cl_2]$ etc.

Formation of tetracyanonickel(II) ion i.e. $[Ni(CN)_4]^{2-}$ ion:

$[Ni(CN)_4]^{2-}$ ion is a complex ion formed by the combination of Ni^{2+} and CN^- ions as

$$Ni^{2+} + 4\,CN^- \longrightarrow [Ni(CN)_4]^{2-}$$

Each CN^- ion is linked to Ni^{2+} ion by means of a co-ordinate bond, where CN^- ion donates lone pair of electrons into empty dsp^2 hybrid orbitals of Ni^{2+} ion. Thus, CN^- ion acts as a donor while Ni^{2+} ion acts as acceptor. Nickel atom in ground state has electron configuration as [Ar] $4s^2$, $3d^8$, $4p^0$. The valence electron configuration is $4s^2$, $3d^8$, $4p^0$, but nickel in $[Ni(CN)_4]^{2-}$ is in ionic form (Ni^{2+}) and therefore the valence electron configuration of Ni^{2+} ion is $4s^0$, $3d^8$, $4p^0$.

Nickel has co-ordination number 4 in this complex ion, which is explained on the basis of the hybridization of the vacant 4s, 4p and $3d_{x^2-y^2}$ orbitals to produce four vacant dsp^2 hybrid orbitals. This hybridization is shown with the help of box diagram as below:

		3d					4s	4p		
Nickel atom (in ground state)	=	↑↓	↑↓	↑↓	↑	↑	↑↓			
Nickel ion (Ni^{2+}) (in ground state)	=	↑↓	↑↓	↑↓	↑	↑				

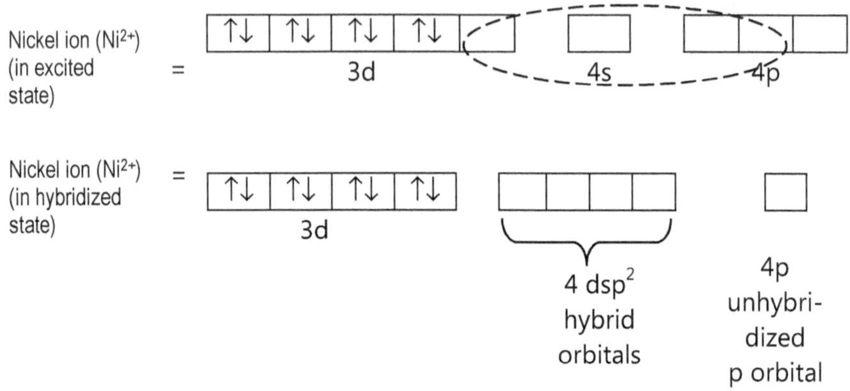

- At the time of bonding, the single electron in $d_{x^2-y^2}$ orbital gets paired with the electron in other d-orbital under the influence of electric field of ligand (CN^-) approaching Ni^{2+} ion. These hybrid orbitals are directed towards the four corners of a regular square in the xy plane making an angle of 90° with one another. dsp^2 hybrid orbitals (vacant) overlap with four completely filled 2p orbitals of four cyanide (CN^-) ions, to form four Ni-CN sigma bonds. The bonds formed are also called as dative bond. All four Ni-CN bonds are identical in every respect and thus give a perfect square planar shape to $[Ni(CN)_4]^{2-}$ ion. Formation of dsp^2 hybrid orbitals [Fig. 3.12 (a)] and formation of $[Ni(CN)_4]^{2-}$ ion [Fig. 3.12 (b)] are shown below:

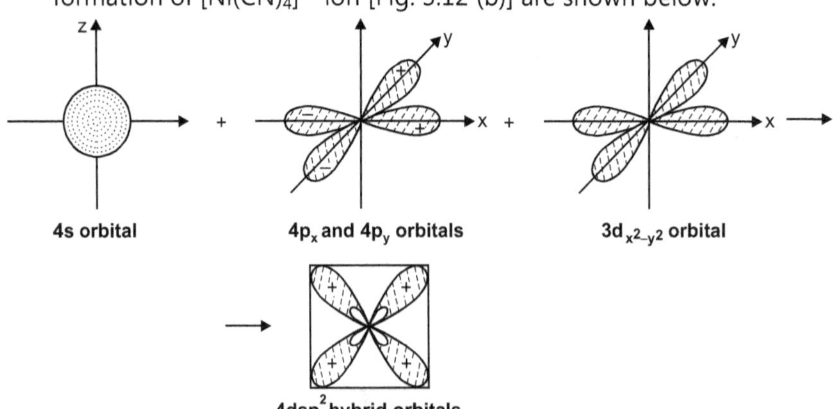

(a) Formation of dsp^2 hybrid orbitals

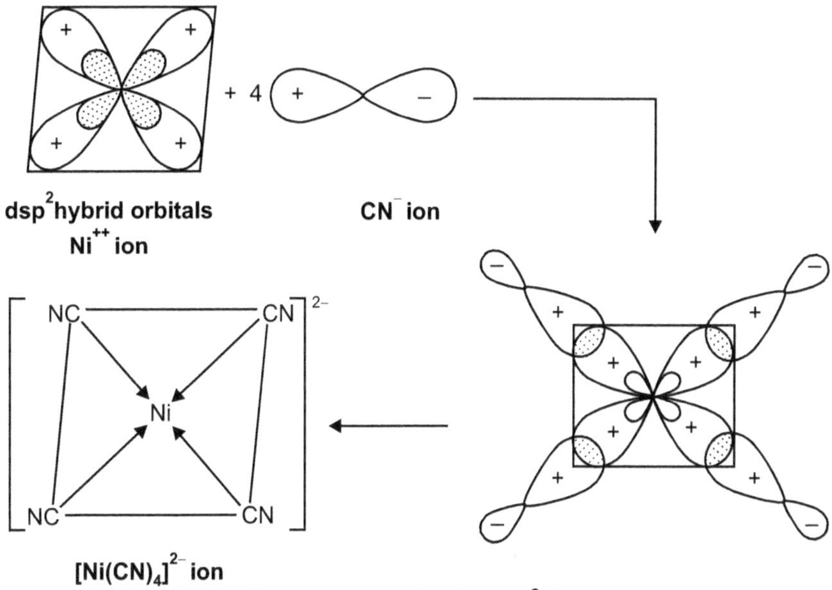

(b) Formation of $[Ni(CN)_4]^{2-}$ ion

Fig. 3.12

(b) dsp^3 Hybridization (Trigonal Bipyramidal Hybridization)

- The intermixing of one s, three p and one d orbitals to form a non-equivalent set of five hybrid orbitals is called dsp^3 hybridization. Since trigonal bipyramidal shape is obtained in this hybridization, it is also called trigonal bipyramidal hybridization.
- When the s, p and d orbitals are from the same shell, i.e. same principal quantum number (n), it is called sp^3d hybridization. But when the d orbital used in process is from penultimate shell (i.e. n – 1), then it is called as dsp^3 hybridization. These hybrid orbitals are directed towards the vertices of a trigonal bipyramid. All the five hybrid orbitals in this are not of same type, but they can be divided into two non-equivalent sets.
  - **(i) Equitorial hybrid orbitals:** This set is supposed to be formed by the hybridization of one s and two p orbitals (sp^2 hybrid orbitals). This set consists of three equivalent hybrid orbitals which are oriented in a plane at the corners of triangle, making an angle of 120°.
  - **(ii) Axial hybrid orbitals:** This set is supposed to be formed by the hybridization of one p (p_z) orbital and one d (dz^2) orbital (dp hybrid orbitals). This set consists of two equivalent orbitals which make an angle of 180° to each other but make an angle

of 90^0 with the equatorial hybrid orbitals situated in triangular plane. Examples of the molecules in which central atom shows dsp^3 hybridization in $Fe(CO)_5$ and sp^3d hybridization in PCl_5, PBr_5, etc.

Formation of [Fe(CO)$_5$] molecule:

- In this complex, central metal atom is iron. Its atomic number is 26 and has valence electron configuration as $4s^2$, $3d^6$. Oxidation number of iron is zero. To form a complex with five carbonyl ligands, five vacant orbitals are required. Iron atom makes it available as shown below:

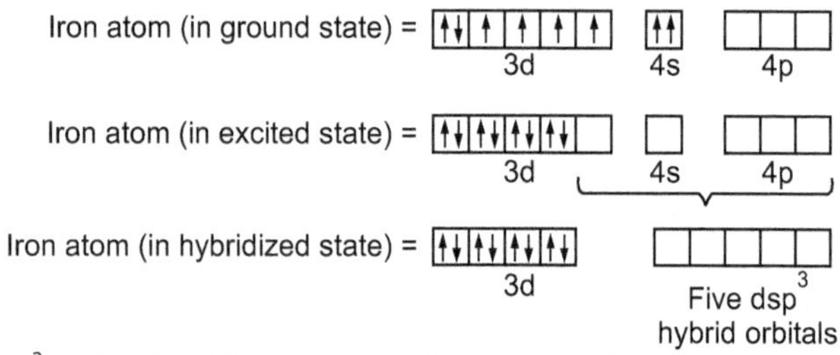

dsp^3 hybrid orbitals orient themselves into two sets, equatorial and axial. Out of five dsp^3 hybrid orbitals, three are in one plane and two are perpendicular to the plane. Formation of dsp^3 hybrid orbitals is shown below in Fig. 3.13.

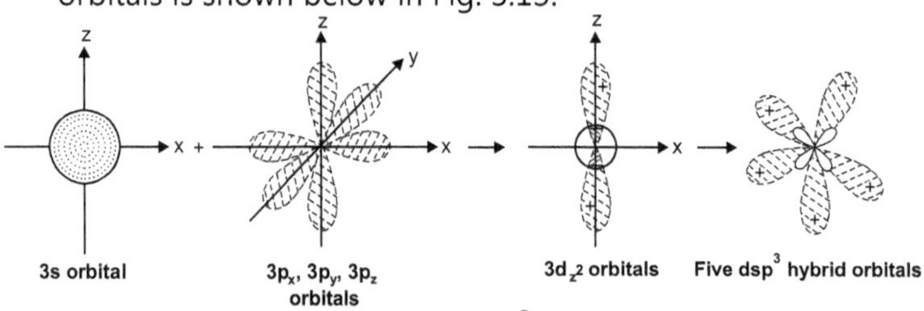

Fig. 3.13: Formation of sp^{3}d hybrid orbitals

dsp^3 hybrid orbitals on iron atom overlap with filled orbital on carbonyl to form five co-ordinate covalent bonds. Fig. 3.14 shows the geometry of [Fe(CO)$_5$] molecule which is trigonal bipyramidal.

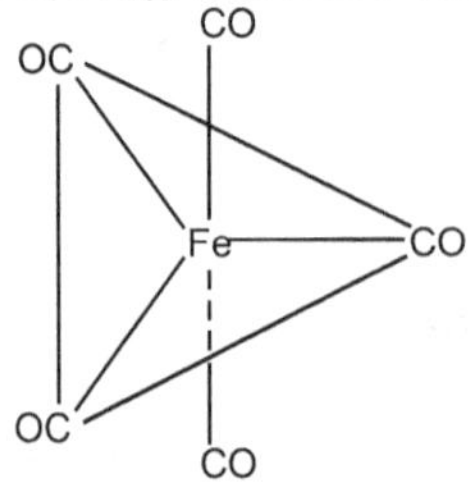

Fig. 3.14: Shape of [Fe(CO)$_5$] molecule

(c) sp^3d^2 Hybridization (Octahedral Hybridization)

- The intermixing of one s, three p and two d orbitals to form a set of six hybrid orbitals is called as sp^3d^2 hybridization. Since octahedral shape is adopted by molecule in sp^3d^2 hybridization, it is also called as octahedral hybridization. Four of the six sp^3d^2 hybrid orbitals are directed towards the corners of a regular square making an angle of 90^0 with one another. The remaining two are perpendicular to the plane of four hybrid orbitals (square). The best example of this type of hybridization is formation of SF_6 molecule.

  Other examples include $[Cr(H_2O)_6]^{3+}$, $[Zn(NH_3)_6]^{2+}$, $[Co(F)_6]^{2+}$ etc.

Formation of $[Cr(H_2O)_6]^{3+}$ complex ion:

- In this complex ion, central metal atom is chromium. Its atomic number is 24 and has valence electron configuration as $4s^2$, $3d^4$. In the complex ion, oxidation number of chromium is +3 and therefore its valence electron configuration is $4s^0$, $3d^3$. Thus chromium ion has incompletely filled 3d orbital and empty 4s, 4p and 4d orbitals. Chromium ion therefore undergoes sp^3d^2 hybridization. Here outer 'd' orbital is used rather than inner 'd' orbital. This is the best example of outer orbital complex. This hybridization can be shown with the help of box diagram as below:

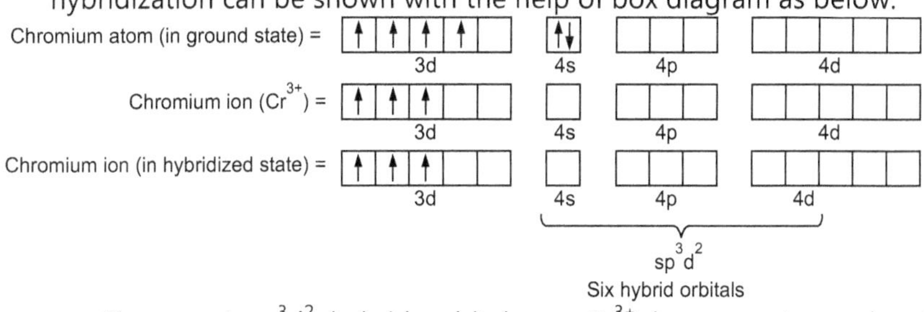

- Six vacant sp^3d^2 hybrid orbitals on Cr^{3+} ion accept a pair of electron from H_2O molecule to form six co-ordinate bonds. The octahedral shape of the resulting complex ion is shown in Fig. 3.15.

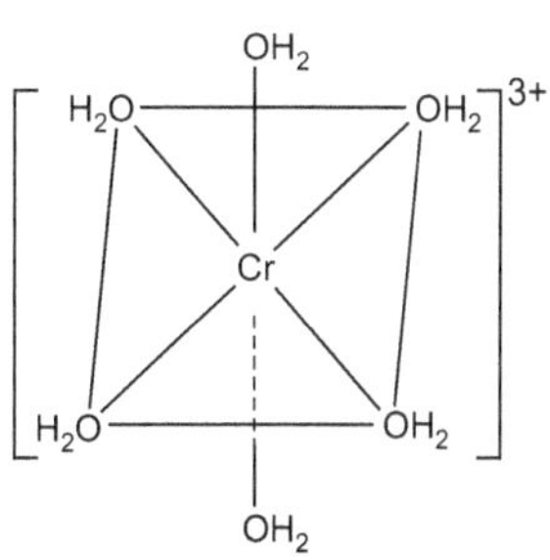

Fig. 3.15: Octahedral geometry of $[Cr(H_2O)_6]^{3+}$ ion

(d) sp^3d^3 Hybridization (Pentagonal Bipyramidal Hybridization)

- The intermixing of one s, three p and three d orbitals to form a set of seven hybrid orbitals is called as sp^3d^3 hybridization. Since pentagonal bipyramidal shape is adopted by molecule in sp^3d^3 hybridization, it is also called as pentagonal bipyramidal hybridization. All the seven hybrid orbitals are not equivalent. Five of these are directed towards the vertices of a regular pentagon making an angle of 72° with one another. The remaining two are perpendicular to the plane of five hybrid orbitals (pentagon). Thus, the bond angles are of two types viz. 72° and 90°.
- **Formation of IF_7 molecule:** In IF_7 molecule, the central atom is iodine. Its ground state electron configuration is [Kr] $4d^{10}$ $5s^2$ $5p^5$. The valence electron configuration of iodine atom in ground state is $5s^2$, $5p^5$. There is only one unpaired electron, but iodine is heptavalent in IF_7 molecule. This can be explained on the basis of promotion of one electron from 5s orbital and two electrons from 5p orbital to the vacant 5d orbital. Thus, in excited state, iodine atom contains seven unpaired electrons. These seven orbitals hybridize to give seven sp^3d^3 type hybrid orbitals. This hybridization is shown with the help of box diagram as below :

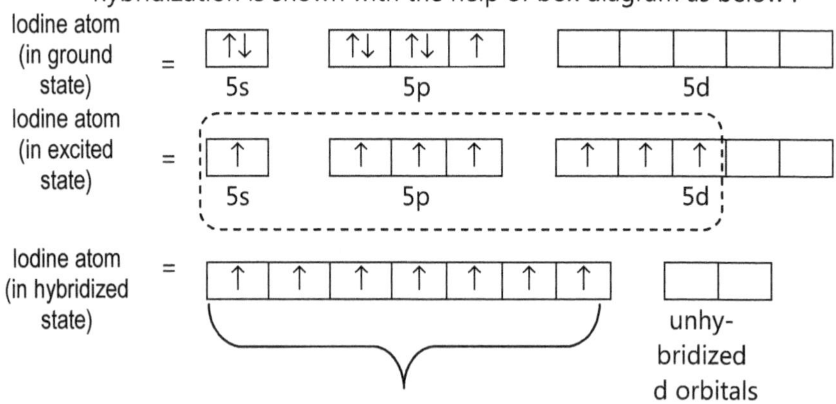

sp^3d^3 hybrid orbitals

sp^3d^3 hybrid orbitals are directed towards the vertices of a pentagonal bipyramid.

- The half filled sp^3d^3 hybrid orbitals overlap with half filled p orbitals of seven fluorine atoms to form seven covalent bonds between iodine and fluorine atoms. In IF_7 molecule, five I-F bonds are equal but are shorter than the two I-F distances of axial fluorine atoms, which in turn are equal in length. This clearly confirms the non-equivalence of the seven sp^3d^3 hybrid orbitals. The pentagonal bipyramidal shape exhibited by IF_7 molecule is shown in Fig. 3.16.

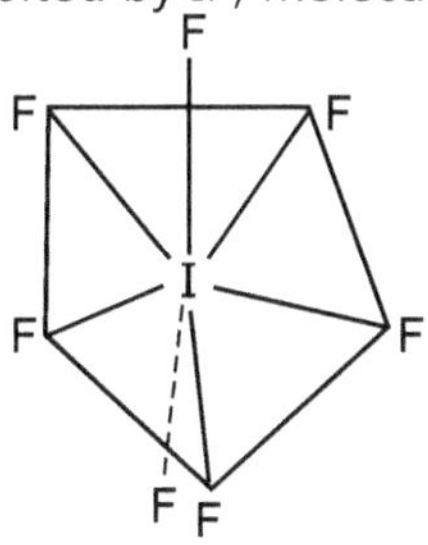

Fig. 3.16: Pentagonal bipyramidal structure of IF_7 molecule

- To conclude the discussion of hybridization concept, it may be stated that, hybridization and the mixing of orbitals is a most useful concept. For effective mixing, the energy of the orbitals must be nearly the same. It is common misconception that hybridization is the cause of a particular molecular shape. This is not so. The reason why any particular shape is adopted is its energy. It should also be noted that the hybridized state is a theoretical step in going from atom to a molecule. The energy of hybrid orbitals cannot be measured and can only be estimated theoretically.
- To remember various types of hybridization, the orbitals involved, shapes of the molecules and bond angles involved, following Table 3.4 may be very useful.

Table 3.4: Different types of hybridization

Sr. No.	Number and type of orbitals	Type of hybridization	Shape of molecule	Bond angle	Examples
1.	2, s and p	Sp	Linear	180°	BeH_2, BeF_2, CO_2.
2.	3, s and p	sp^2	Trigonal planar	120°	BF_3, BCl_3
3.	4, s and p	sp^3	Tetrahedral	109°.28'	CH_4, CCl_4 $[Mn(Cl)_4]^{2-}$
4.	4, s, p and d	dsp^2	Square planar	90°	$[Ni(CN)_4]^{2-}$, $[Pt(Cl)_4]^{2-}$
5.	5, s, p and d	dsp^3	Trigonal bipyramid	120° and 90°	$Fe(CO)_5$
6.	6, s, p and d	sp^3d^2	Octahedral	90°	$[Cr(H_2O)_6]^{3+}$
7.	7, s, p and d	sp^3d^3	Pentagonal bipyramidal	72° and 90°	IF_7

3.13 THE VALENCE SHELL ELECTRON PAIR REPULSION (VSEPR) THEORY

3.13.1 Introduction

- Molecular geometry or shape of molecule is generally due to three-dimensional arrangement of atoms or ions in a molecule. Many physical and chemical properties, such as melting and boiling points, density, hardness or softness, various chemical reactions that molecule undergo, are directly related to molecule's geometry. To understand these properties clearly, knowledge of molecular geometry is must. In general, bond lengths and bond angles must be determined experimentally. However, there is simple procedure to predict the overall geometry of a molecule. If we know the number of electrons surrounding a central atom, then we may predict successfully the geometry of a molecule.
- The basis of this approach is the assumption that, the electron pairs in the valence shell of an atom repel one another. (The valence shell is the outermost shell of an atom, that holds the electrons which are usually involved in bonding). In a covalent bond formation, a pair of electrons (called as bonding pair) is responsible for holding two atoms together. However, in a polyatomic molecule, where there are two or more bonds between the central atom and the other atoms, the repulsion between the electrons in different bonding pairs causes them to remain as far apart as possible. The geometry that molecule finally adopts minimizes this repulsion. Therefore, this approach to the study of molecular geometry is called as valence shell electron pair repulsion (VSEPR) model and the theory as VSEPR theory.

3.13.2 Need of the VSEPR Theory

- In the molecules of the type AB_x like $BeCl_2$, BF_3, CH_4 etc. the central atoms namely Be, B and C have two, three and four electron pairs around them respectively. All these electron pairs are bonded pairs of electrons, that is, there are no lone pairs present. In such molecules, the maximum repulsion between the bonded pairs gives linear, triangular and tetrahedral geometries to the molecules respectively. These geometries are said to be regular (expected) geometries of molecules. However, in molecules like NH_3 and H_2O, the central atoms N and O are also surrounded by four electron pairs but all are not bonded pairs. Since there are four electron pairs around N and O, it is expected that these molecules must be tetrahedral in shape and also should have the bond angle as 109° 28' like that of CH_4 molecule. However, these molecules have a different shape like NH_3 molecule is pyramidal in shape with a bond angle (NHN) 107° 48' and H_2O has inverted V shape with a bond angle (HOH) 104° 27'.

- This little different behaviour of some molecules, i.e. the unexpected geometries adopted by molecules need to be explain. In order to explain the irregular (unexpected) geometries and bond angles of many molecules, the VSEPR theory was proposed. The theory was given by Gillespie and Nyholm. This theory considers the effect of both bonded pairs as well as lone pairs of electrons on the shape and bond angles of the molecules.

3.13.3 Assumptions of VSEPR Theory

(i) In a molecule, electrons in the valence shell of the central atom (say A) form pairs with opposite spins.

(ii) Electron pair around the central atom usually adjust themselves with respect to one another in such a way that the force of repulsion among them is minimum and thus the stability to the molecule is maximum.

(iii) The pairs which are used in bond formation are called bonded pairs (B.P.) while the pairs which are not used in bond formation are called lone pairs (L.P.).

(iv) A lone pair requires more space, i.e. it occupies greater volume than the bonded pair because the lone pair of electrons is under the influence of only one nucleus while bonded pair is under the influence of two nuclei.

(v) Since lone pair have bigger volume, it repels the other pairs of electrons to a greater extend. Thus, the repulsive interactions between two lone pairs of electrons is more than between lone pair and bonded pair, which in turn is greater than between two bonded pairs. Thus we have

L.P. – L.P. repulsion > L.P. – B.P. repulsion > B.P. – B.P. repulsion.

(vi) When the central atom in a molecule is surrounded by only bonded pairs, the repulsive forces between electron pairs are similar (equivalent) and therefore molecular geometry is regular.

(vii) When the central atom in a molecular is surrounded by both bonded pairs and lone pairs, the repulsive forces are not similar (non-equivalent) and therefore the geometry is irregular or distorted.

3.13.4 Applications of VSEPR Theory to Explain Geometry of Molecules (April 18, Oct. 16)

- The factors that govern the geometry of a molecule are :
  - (i) Total number of electron pairs in the valence shell of the central atom.
  - (ii) Total number of lone pairs in the molecule.
  - (iii) Availability of the low energy orbitals.
- We know that if two pairs of electrons are present they are arranged linearly, for example, BeF_2, if three electron pairs are present then, molecule is trigonal planer, for example, BF_3 and if four electron pairs are present then, molecule is tetrahedral, for example CH_4. On the similar lines five pairs should give a trigonal bipyramid and six pairs give an octahedral geometry. All these molecules have their general formula AB_X, where, x = 2 to 6. In such molecules, regular geometries are obtained when only bonded electron pairs are in surrounding of central atom. But, if the central atom has one or more lone pairs of electrons, the geometry of the molecule becomes irregular and the bond angles usually do not agree with the expected or theoretical one.
- VSEPR theory exactly finds use in explaining irregular geometries and bond angles of the molecules. With the help of some examples, we will see how VSEPR theory is useful and we will try to understand the major role played by lone pair of electrons to decide the geometry of molecule.

(i) Bonding and shape of ClF_3 molecule (AB_3E_2): ClF_3 molecule is of the type AB_3E_2, where A is the central atom, B is the atom of other element and E represents the number of lone pairs on the central atom. ClF_3 contains three bonded pairs and two lone pairs. Chlorine atom undergoes **sp^3d** hybridization in which one 3s, three 3p and one 3d orbitals combine to produce five sp^3d hybrid orbitals. Out of these five hybrid orbitals, two orbitals contain lone pairs of electrons and the remaining three are half filled. These half filled sp^3d hybrid orbitals of chlorine atom overlap co-axially with three half filled p-orbitals of three fluorine atoms. Three different possible arrangements are shown in Fig. 3.17.

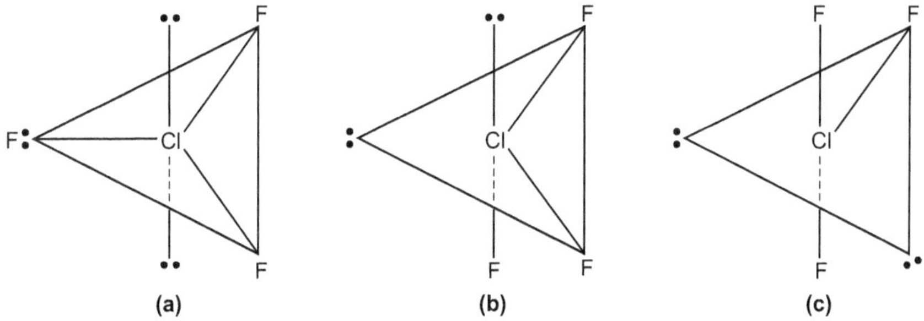

Fig. 3.17: Possible structures for chlorine trifluoride

A structural study of ClF_3 shows that the molecule is T shaped, with bond angle of 87° 40'. This is close to 90° and suggests structure (c). The distortion from 90° is because of repulsion between the lone pairs. Two bond lengths are the same and are different from the third. This is because a trigonal bipyramid is not a regular shape. Equatorial bonds (those in the triangle) are different from apical bonds (those pointing up and down). The X-ray structure of ClF_3 shows that the molecule is T shaped with a bond angle 87° 40' and bond lengths 1.698 A° and 1.598 A°. Fig. 3.18 shows the structure of ClF_3 molecule.

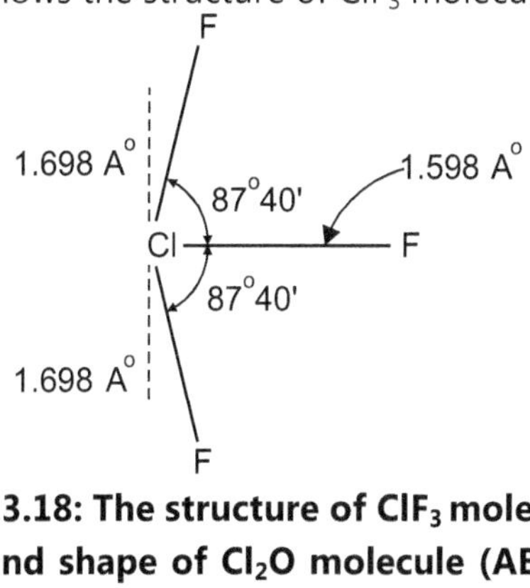

Fig. 3.18: The structure of ClF_3 molecule

(ii) Bonding and shape of Cl_2O molecule (AB_2E_2): Cl_2O molecule is of the type AB_2E_2, where A is the oxygen atom, B is the chlorine atom and E is the number of lone pairs on the oxygen atom. Cl_2O thus contain two bonded pairs and two lone pairs. Oxygen is the central atom in Cl_2O, it has the electron configuration $1s^2$, $2s^2$, $2p^4$. It is clear that it has two half filled 2p orbitals which can form two covalent bonds with two chlorine atoms. It is not necessary therefore to excite the oxygen atom. Since p orbitals are perpendicular to each other, Cl_2O molecule should be angular and the bond angle must be 90°. Actually Cl_2O molecule is inverted 'V' shape with Cl-O-Cl bond angle 111°.

- This can be explained by assuming that oxygen atom undergoes sp^3 hybridization using its filled 2s and one 2p orbitals and two half filled 2p orbitals. It produces **four sp^3 hybrid orbitals** out of which two contain lone pair of electrons while two are half filled. The two half filled sp^3 hybrid orbitals are used to form bonds with two chlorine atoms. The expected shape of the molecule is tetrahedral with Cl-O-Cl bond angle 109° 28'. Actually we expect decrease in bond angle due to strong repulsion between two lone pairs as seen in the case of F_2O (F-O-F bond angle is 102°). Here in the case of Cl_2O, Cl-O-Cl bond angle is 111° and it is because of much bulkier size of Cl atom compared to F atom. The repulsion between two chlorine atoms is predominant than repulsion between two lone pairs and hence the bond angle is found to increase. Fig. 3.19 shows the structure of Cl_2O molecule.

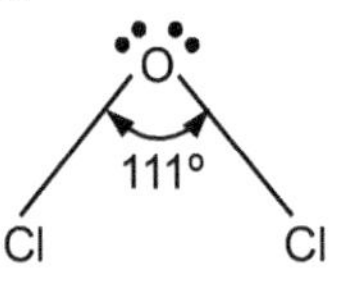

Fig. 3.19: The geometry of Cl_2O molecule

(iii) Bonding and shape of BrF_5 molecule (AB_5E) :

- BrF_5 molecule is of the type AB_5E. BrF_5 contains five bonded pairs and one lone pair. Central bromine atom undergoes **sp^3d^2 hybridization** in which one 4s, three 4p and two 4d orbitals combine to produce six sp^3d^2 hybrid orbitals. Out of these five hybrid orbitals contain one electron each and sixth orbital contains lone pair of electrons. Five half filled **sp^3d^2** hybrid orbitals of bromine atom overlap with half filled p orbitals of five fluorine atoms. This hybridization leads to square-based pyramidal structure of molecule that is octahedral with one position unoccupied. Five unpaired electrons form bonds with five fluorine atoms, plus one lone pair giving a total of six electron pairs. Structure is octahedral with one position occupied by a lone pair. It is also called square-based pyramidal. Fig. 3.20 shows the structure of BrF_5 molecule.

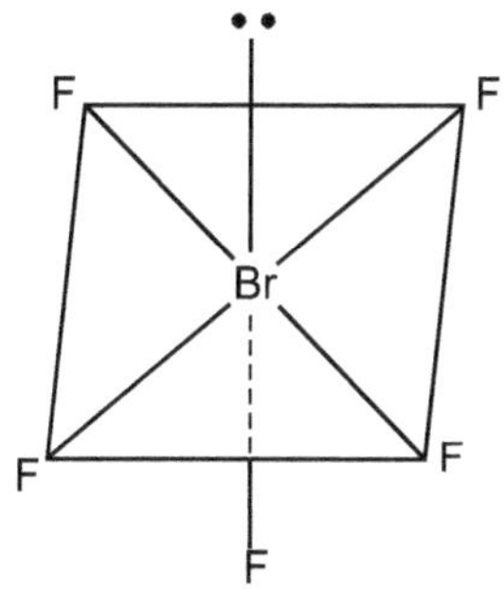

Fig. 3.20: Structure of BrF_5 molecule

- To summarize the VSEPR theory, it may be stated that theory provides reliable predictions of the geometries of variety of molecular structures. Many chemists accept and use this VSEPR approach because it is simple and has lot of utility. Although some limitations or shortcomings are there it is safe to say that, such an assumption leads to useful predictions. Table 3.5 may be useful to remember various molecular structures.

(iv) Bonding and Shape of XeO_3 molecule (AB_3E):

- XeO_3 molecule is of the type AB_3E. This molecule contain six bonded pairs and one lone pair. Bonding in this molecule is explained using box diagram as below:

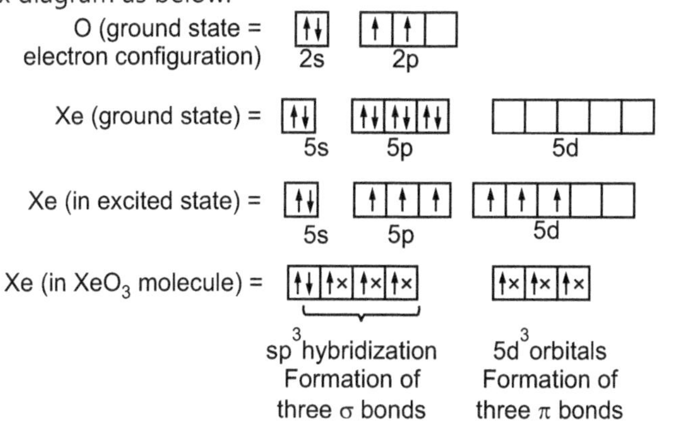

The cross indicate electrons of oxygen atoms

- Xenon atom in excited state undergoes sp^3 hybridization. Three unpaired electrons in **sp^3 hybrid orbitals** are involved in three sigma bonds with three oxygen atoms. The remaining three unpaired electrons in 5d orbital are involved in π bond formation. There is one lone pair of electron which occupies the 4th position of tetrahedron and thus reduce the geometry from tetrahedron to pyramidal (tetrahedron with one corner unoccupied). Out of two unpaired electrons on oxygen atom, one electron is involved in σ bond formation with xenon and other unpaired electron in π bond formation. Fig. 3.21 shows the structure of XeO_3 molecule.

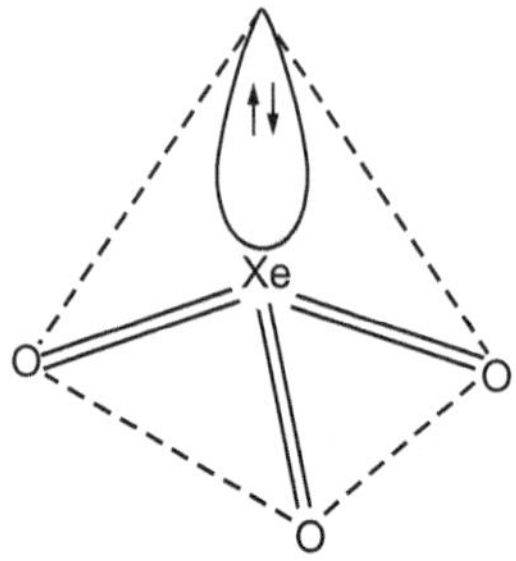

Fig. 3.21: Structure of XeO_3 molecule

(v) Bonding and shape of $XeOF_4$ molecule (ABC_4E):

- $XeOF_4$ molecule is of the type ABC_4E. This molecule contain six bonded pairs and one lone pair. Bonding in this molecule is explained using box diagram as below:

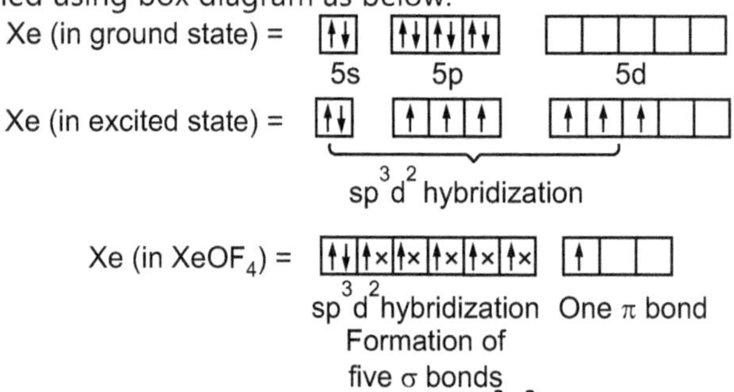

- In $XeOF_4$ molecule, xenon undergoes **sp^3d^2 hybridization**. There is one lone pair of electron in one of the sp^3d^2 hybrid orbital. Five unpaired electrons in five sp^3d^2 hybrid orbitals of xenon are involved in five sigma bond formation. (Four sigma bonds with four fluorine atoms and fifth sigma bond with oxygen atom). The remaining one unpaired electron in unhybridized 5d orbital is involved in pi bond formation. Fig. 3.22 shows the structure of $XeOF_4$ molecule.

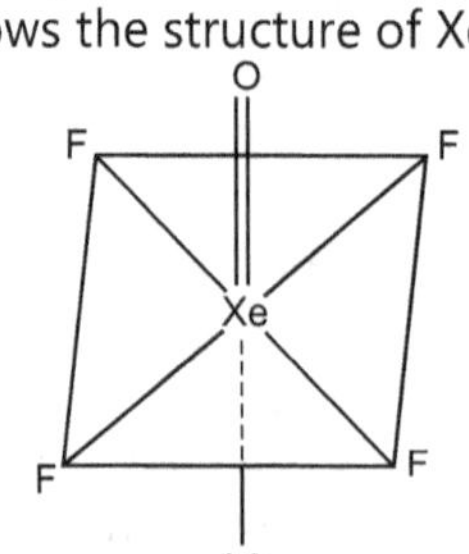

Fig. 3.22: Structure of $XeOF_4$ molecule

- There is one lone pair of electron which occupies the 6th position of tetrahedron and thus the geometry of the molecule is square pyramidal rather than octahedral.

Table 3.5: The effects of bonding and lone pairs on bond angles and shapes of some common molecules

Molecular formula	Number of bonded pairs	Number of lone pairs	Shape of molecule		Bond angle	
			Expected	Actual	Expected	Actual
$BeCl_2$	2	0	Linear	Linear	180°	180°
BF_3	3	0	Plane triangle	Plane triangle	120°	120°
CH_4	4	0	Tetrahedral	Tetrahedral	109° 28'	109° 28'
NH_3	3	1	Tetrahedral	Pyramidal	109° 28'	107° 48'
NF_3	3	1	Tetrahedral	Pyramidal	109° 28'	102° 30'
H_2O	2	2	Tetrahedral	Inverted V	109° 28'	104° 27'
F_2O	2	2	Tetrahydral	Inverted V	109° 28'	102°
PCl_5	5	0	Trigonal bipyramid	Trigonal bipyramid	120° and 90°	120° and 90°
ClF_3	3	2	Trigonal bipyramid	T-shaped	90°	87° 40'
BrF_3	3	2	Trigonal bipyramid	T-shaped	90°	86° 12'
ICl_2^-	2	3	Trigonal bipyramidal	Linear	180°	180°
SF_6	6	0	Octahedral	Octahedral	90°	90°
BrF_5	5	1	Octahedral	Square pyramidal	90°	84° 30'
Cl_2O	2	2	Tetrahedral	Inverted V	109° 28'	111°
$[Cr(H_2O)_6]^{3+}$	6	0	Octahedral	Octahedral	90°	90°

SUMMARY

- There are four types of bonds namely, ionic, covalent, coordinate and metallic bond.
- Ionic bond is formed when electropositive element react with electronegative element. Here a complete transfer of one or more electrons of one atom to the another dissimilar atom is possible.
- The bond formed by the mutual sharing of electron pairs between the atoms of the same or different elements is called a covalent bond or electron pair bond.
- Sometimes more than two electrons are shared between a pair of atoms. If four electrons are shared then there are two bonds and this arrangement is called double bond. If six electrons are shared then there are three bonds, and this is called a triple bond.
- A special type of covalent bond in which both the electrons required for sharing between two atoms are contributed by one atom only is called co-ordinate bond.
- The process of mixing and redistribution of atomic orbitals of different energies in an isolated atom to form a new set of orbitals of equal energies is called hybridization.
- There are different types of hybridization. Atomic orbitals of suitable energy and symmetry combine to form hybrid orbitals. Some examples are sp, sp^2, sp^3, dsp^2, dsp^3, d^2sp^3, sp^3d^2, d^3sp^2 etc.
- According to VSEPR theory, geometry of the molecule is affected by presence of lone pairs in the molecule.

PRACTICE QUESTIONS

(A) Multiple Choice Questions (MCQs) :

1. Ionic bond is possible by following combination ______
   (a) electropositive element + electropositive element
   (b) electronegative element + electronegative element
   (c) electropositive element + electronegative element
   (d) none of the above

2. Covalent bond is formed by ______
   (a) loss of electron(s) (b) gain of electron(s)
   (c) sharing of electron(s) (d) none of these
3. In covalency _____
   (a) the transference of electrons takes place
   (b) sharing of electrons takes place
   (c) the electrons are shared by only one atom
   (d) none of these
4. In a double bond connecting two atoms, there is a sharing of ____
   (a) 2 electrons (b) 4 electrons
   (c) 1 electron (d) 5 electrons
5. How many π bonds are there in nitrogen molecule?
   (a) 0 (b) 1
   (c) 2 (d) 3
6. Which of the following contain a co-ordinate covalent bond?
   (a) NH_4^+ (b) HCl
   (c) H_2O (d) CCl_4
7. Polar compounds are soluble in ____
   (a) concentrated acids (b) organic solvents
   (c) polar solvents (d) non-polar solvents
8. Which one of the following molecules show non-polar covalent bonding?
   (a) H_2 (b) HCl
   (c) HF (d) H_2O
9. Which one of the following molecules show covalent bonding?
   (a) H_2 (b) NaCl
   (c) F_2 (d) CH_4
10. Elements which have more electronegativity difference tend to form ____
    (a) covalent bond (b) metallic bond
    (c) ionic bond (d) hydrogen bond

11. sp^3 hybridization is due to intermixing of
    (a) one s and 3p orbitals (b) one s, 3p and 1d orbitals
    (c) all s and p orbitals (d) all s, p and d orbitals
12. Each hybrid orbital can accommodate maximum electrons ______
    (a) 4 (b) 2
    (c) 3 (d) 1
13. Shape of the molecule involving sp^3d hybridization is ______
    (a) triangular (b) linear
    (c) trigonal bipyramidal (d) tetrahedral
14. Which one of the following molecules is octahedral ?
    (a) BrF_5 (b) PCl_5
    (c) IF_7 (d) SF_6
15. Hybridization used in the formation of IF_7 molecule is ______
    (a) sp^3d^3 (b) sp^3d^2
    (c) sp^3d (d) d^2sp^3
16. Bond angle between ClOCl in ClO_2 is ______
    (a) 108° 28' (b) 107° 48'
    (c) 111 (d) 102° 30'
17. The correct order of magnitude of repulsion is ______
    (a) BP – BP > LP – BP > LP – LP
    (b) LP – BP > LP – LP > BP – BP
    (c) LP – LP > LP – BP > BP – BP
    (d) LP – BP > BP – BP > LP – LP
18. Shape of BrF_5 molecule is ______
    (a) trigonal bipyramidal (b) pyramidal
    (c) distorted tetrahedron (d) square pyramidal
19. ClF_3 molecule is ______
    (a) T-shaped (b) pentagonal
    (c) trigonal bipyramidal (d) square pyramidal
20. $[Cr(H_2O)_6]^{3+}$ is ______
    (a) triangular (b) linear
    (c) pyramidal (d) octahedral

ANSWERS

1. - (b)	2. - (c)	3. - (b)	4. - (b)	5. - (c)
6. - (a)	7. - (c)	8. - (a)	9. - (c)	10. - (c)
11. - (a)	12. - (b)	13. - (d)	14. - (d)	15. - (a)
16. - (c)	17. - (c)	18. - (a)	19. - (a)	20. - (d)

(B) State True or False Questions :

1. Ionic compounds have no definite shape because ionic bond is non-directional.
2. Co-ordinate covalent bond is present in NH_4^+.
3. Water molecule is non-polar.
4. Linear overlap of p-orbitals leads to formation of π-bond.
5. N_2 molecule shows double bond.
6. In oxygen molecule, there is oxygen-oxygen triple bond.
7. In nitrogen molecule, there is presence of one σ-bond and two π-bonds.
8. Covalent compounds are soluble in non-polar solvents.
9. Ionic compounds are soluble in water.
10. In covalent compounds, there is sharing of electrons.
11. π-bond is weaker than σ-bond.
12. Those orbitals can take part in bonding which contain unpaired electrons.
13. The process of mixing and redistribution of atomic orbitals of different energies in an isolated atom to form new set of orbitals of equal energies is called hybridization.
14. Hybridization using s, p and d orbitals is not possible.
15. sp^3 hybridization is also called trigonal hybridization.
16. Shape of the molecule after dsp^2 hybridization is square planar.
17. Hybridization involved in the formation of Cl_2O molecule is sp^2.
18. IF_7 molecule is distorted octahedron.
19. Position of lone pair in molecule decides the stability of the molecule.
20. Tetrahedral bond angle is 109° 28'.
21. $TeCl_4$ molecule is sometimes described as distorted tetrahedron.

ANSWERS

1. - (T)	2. - (T)	3. - (F)	4. - (F)	5. - (F)
6. - (F)	7. - (T)	8. - (T)	9. - (T)	10. - (T)
11. - (T)	12. - (F)	13. - (T)	14. - (F)	15. - (F)
16. - (T)	17. - (F)	18. - (F)	19. - (T)	20. - (T)
21. - (T)				

(C) Short and Long Answer Questions :

1. What is octet rule? How does it lead to the formation of electrovalent and covalent bond?
2. Define and explain chemical bond. How do atoms combine to form compounds?
3. Define (i) Ionic bond, (ii) Covalent bond, (iii) Co-ordinate bond and (iv) Metallic bond.
4. Explain with the help of suitable example, formation of ionic bond and covalent bond.
5. Explain the terms (i) Co-ordinate bond and (ii) Metallic bond with suitable example.
6. Define : (i) Co-ordinate bond, (ii) Sigma bond.
7. Discuss the formation of co-ordinate bond with suitable examples.
8. What are ionic and covalent bonds? How are they formed?
9. With the help of neat diagram, explain in detail what is metallic bond.
10. What are the limitations of octet theory?
11. What are the postulates of valence bond theory?
12. Give the important points of Heitler-London theory.
13. State the postulates of Pauling-Slater theory.
14. Describe with the help of potential energy curve the formation of H_2 molecule on the basis of valence bond theory.
15. What is atomic orbital overlap? Give the factors which affect the extent of overlap.

16. Give the points of comparison between sigma bond and pi bond. Explain why sigma bond is stronger than pi bond.
17. What do you understand by polar and non-polar character of covalent bonds?
18. What are the different types of overlap between s, p and d orbital ? Give one example of each.
19. Explain the formation of H_2, F_2, HF molecules on the basis of atomic orbital overlap.
20. Explain the formation of $p\pi$-$d\pi$ by p-d overlapping with the help of suitable example.
21. Explain the formation of HF molecule on the basis of valence bond theory.
22. Explain the terms 'single bond' and 'double bond'.
23. Compare between 'ionic bond' and 'covalent bond'.
24. What is hybridization? What is the need of hybridization?
25. Define the terms hybridization and hybrid orbitals.
26. What are important features of hybridization and hybrid orbitals?
27. What are uses of hybridization?
28. State the different types of hybridization using s, p and d orbitals.
29. Describe the steps involved in hybridization of atomic orbitals.
30. What is diagonal hybridization? Describe the shape of BeH_2 molecule on the basis of sp hybridization.
31. Explain with suitable example sp^2 hybridization. Why it is called trigonal hybridization?
32. What is tetrahedral hybridization? Explain the formation and shape of $[MnCl_4]^{2-}$ molecule with the help of sp^3 hybridization.
33. Explain the formation of $[Ni(CN)_4]^{2-}$ ion with the help of dsp^2 hybridization.
34. What is trigonal bipyramidal hybridization? Explain dsp^3 hybridization with the help of suitable example.
35. What is octahedral hybridization? Explain the formation of $[Cr(H_2O)_6]^{3+}$ molecule on the basis of sp^3d^2 hybridization.

36. Explain sp^3d^3 hybridization. Why it is called pentagonal bipyramidal hybridization?
37. With the help of suitable example, explain what is the difference between dsp^3 and sp^3d hybridization?
38. State the type of hybridization and predict the shape of the following molecules : PCl_5, BrF_5, BCl_3, I_3^-, $[Pt(Cl)_4]^{2-}$.
39. What do you mean by hybridization? How would you account for the tetravalency of carbon with the help of methane molecule?
40. Explain the bonding and shapes of following molecules :
    NF_3, PBr_5, BF_3, $[Ni(CO)_4]^{2-}$ ion.
41. What is the basis of VSEPR theory? Which electron pairs are considered by this theory?
42. Explain the need of VSEPR theory.
43. What are the assumptions of VSEPR theory?
44. How do lone pair of electrons on the central atom affect the geometrical arrangement of other atoms in a molecule? Explain the interactions between lone pair and bonded pairs of electrons.
45. On the basis of VSEPR theory, explain the shapes and bond angles in following molecules : (a) ClF_3, (b) Cl_2O, (c) $XeOF_4$, (d) $TeCl_4$, (e) BrF_5, (f) XeO_3.
46. Explain the bonding and shapes of the molecules : ClF_3 and BrF_3 using VSEPR approach.
47. Explain the factors affecting geometry of the molecule.
48. Define lattice energy. What is Born-Haber cycle ? Explain the use of Born-Haber cycle in calculating the lattice energy of sodium chloride.
49. Give important applications of Born-Haber cycle.
50. Explain Born-Haber cycle for following ionic solids : CaO, MgO, $CaCl_2$, CaF_2
51. Give Born-Lande equation to calculate lattice energy of an ionic compound and explain various terms involved in it.

UNIVERSITY QUESTIONS

April 2009

1. Define electrovalent bond and covalent bond.
2. Draw the structures of following molecules:
   ClF_3, $TeCl_4$, BrF_5, OF_2
3. What are the different types of overlap between s and p orbitals? Give suitable examples.
4. State the postulates of Pauling-Slater theory.
5. Explain the bonding and shape of NH_3 and H_2O molecules on the basis of VSEPR theory.

April 2010

1. Define metallic bond and covalent bond.
2. What is hybridization? Give its hypothetical sequence of steps involved in the formation of hybrid orbitals.
3. Define overlap and explain which factors affect the extent of overlap of atomic orbital.
4. Explain the bonding and shape of H_2O and H_2S molecules on the basis of VSEPR theory.
5. State the postulates of Pauling-Slater theory.

Nov. 2010

1. State type of hybridization and shape of the following:
   (i) PCl_5, (ii) IF_7
2. What are the limitations of octet theory?
3. What is dsp^2 hybridization? Explain with suitable example.
4. Sigma bond is stronger than pi bond. Explain with suitable example.
5. Give assumptions of VSEPR theory.
6. Explain the formation of F_2 and N_2 molecules on the basis of atomic orbital overlap.
7. What are similarities of hydrogen with halogens and alkali metals?

Oct. 2011

1. Define covalent bond and ionic bond.
2. State type of hybridization and shape of the following: (i) BF_3 and (ii) SF_6.
3. What is sp^3d hybridization? Explain with suitable example.
4. Explain the process of bond formation on the basis of atomic orbital overlap. Give the factors which affect magnitude of overlap of atomic orbital.
5. Explain the bonding and shape of H_2O and H_2S molecules on the basis of VSEPR theory.
6. State the postulates of Heitler-London theory.
7. What is hybridization? Explain the formation of C-H bond in CH_4 molecule.
8. Explain the bonding and shape of BeF_2 molecule.
9. What is atomic orbital overlap? What are different types of overlap between s and p orbitals? Give one example each.
10. Explain the bonding and shape of NH_3 and ICl_2 on the basis of VSEPR theory.

April 2015

1. Draw the structures of $[Ni(CN)_4]^{2-}$, PCl_5, BF_3 and $[Cr(H_2O)_6]^{2+}$.
2. Explain the shape of BrF_5 molecule on the basis of VSEPR theory.

October 2015

1. Explain atomic orbital overlap with suitable example.
2. What is atomic orbital overlap ? Give factors affecting it.
3. What is sp^3 hybridization ? Explain with suitable example.
4. Draw the structures of XeO_3, $XeOF_3$, BrF_5 and $TeCl_4$.
5. Explain the formation of N_2 molecule on the basis of atomic orbital overlap.

April 2016

1. Explain in brief formation of covalent bond with suitable example.

2. What is hybridization ? Explain the formation of CH_4 molecule.
3. What are the assumptions of VSEPR theory ?
4. State the assumptions of Pauling-Slater theory.
5. Explain the formation of F_2 molecule on the basis of atomic orbital overlap.

October 2016

1. Explain in brief formation of ionic bond with suitable example.
2. What is sp^3 hybridization ? Explain with suitable example.
3. What are the limitations of VSEPR theory ?
4. Explain the formation of O_2 molecule on the basis of atomic orbital overlap.
5. PCl_5 shows dsp^3 hybridization. Explain.

April 2017

1. Explain in brief formation of covalent bond with suitable example.
2. What is sp^2 hybridization ? Explain the formation of BF_3 molecule.
3. Define atomic orbital overlap. Give factors affecting it.
4. Draw the structures of :

   (i) BrF_3, (ii) $TeCl_4$, (iii) XeO_3, (iv) IF_7.
5. Define the terms :

   (i) Co-ordinate bond, (ii) Sigma bond, (iii) Octet rule, (iv) Metallic bond.

October 2017

1. Explain in brief sigma (σ) and pi (π) bond.
2. $[Ni(CN)_4]^{2-}$ shows dsp^2 hybridization. Explain.
3. Draw the structures of :

   (i) ClF_3, (ii) $XeOF_4$, (iii) XeO_3, (iv) PCl_5.
4. Explain bond angle in BrF_3 is less than that in ClF_3.
5. Explain in brief formation of co-ordinate bond with suitable example.

April 2018

1. Explain the terms : Covalent bond and co-ordinate bond.
2. What is sp^3 hybridization ? Explain the formation of NH_3 molecule.
3. Explain the formation of N_2 molecule on the basis of atomic orbital overlap.
4. Draw the structures of :

   (i) BF_3, (ii) IF_7, (iii) $XeOF_4$, (iv) XeO_3
5. Define the terms :

   (i) σ (sigma) bond, (ii) π (Pi) bond, (iii) Normality, (iv) Molarity.

***

www.ingramcontent.com/pod-product-compliance
Lightning Source LLC
LaVergne TN
LVHW021939191026
850499LV00083B/1350

* 9 7 8 9 3 8 9 6 8 6 7 0 8 *